5

TESI

THESES

tesi di perfezionamento in Fisica sostenuta il 20 gennaio 2007

COMMISSIONE GIUDICATRICE
Mario Tosi, Presidente
Rosario Fazio
Bilal Tanatar
Giuseppe La Rocca
Marco Polini
Zehra Akdeniz
Francesco Giazotto

Mohammad Reza Bakhtiari
Nanoscience Center
P.O. Box 35
University of Jyväskylä
40014 Jyväskylä
Finland

Quantum Gases in Quasi-One-Dimensional Arrays

Mohammad Reza Bakhtiari

Quantum Gases in Quasi-One-Dimensional Arrays

EDIZIONI
DELLA
NORMALE

ISBN: 978-88-7642-319-2

to my family

Contents

Summary

This thesis is addressed to two sets of problems in the physics of ultra-cold atomic gases in one-dimensional configurations: (i) transport of matter in various types of linear arrays of potential wells, and (ii) the structure of two-component Fermi gases with unequal spin populations (as usual in atomic physics, the word "spin" is here used to denote two internal degrees of freedom of the atom).

Chapter 1 is an overview of theoretical and experimental facts on ultra-cold atomic gases. We start from the Bose-Einstein theoretical prediction on condensation in the ground state for a bosonic system at very low temperature. The main difficulty in realizing Bose-Einstein condensation (BEC) was reaching ultra-low temperatures and relatively high gaseous densities: finally in 1995 BEC was achieved by means of the laser cooling and evaporative cooling techniques. This achievement paved the way for other research areas like low-dimensional (two and one-dimensional) Bose gases at ultralow temperature which we will mention briefly. We then proceed to fermions: the key difference between fermions and bosons lies in their statistics and Pauli blocking causes the evaporative cooling method to fail when applied to a one-component fermion gas. Fermionic gases show specific behaviors known as Pauli blocking and Fermi pressure. We next specifically focus on two issues for 1D gases: the so-called Confinement Induced Resonance (CIR), which points out the effective coupling constant in a tightly confined 1D boson gas, and the Luttinger-liquid model, which describes 1D interactive fermions in the low momentum-low energy sector.

Magneto-optical traps and optical lattices are described as two key experimental tools in the second part of this chapter. The last part of the chapter presents quantum phase transitions in ultra-cold atomic gases. The main effort in this direction started with the theoretical work by D. Jaksch *et al.* [Phys. Rev. Lett. **81**, 3108 (1998)], who predicted a continuous (zero-temperature) quantum phase transition from a superfluid to a

Mott insulator phase. Seminal experiments by M. Greiner *et al.* [Nature, **415**, 39(2002)] verified this theoretical idea. This experimental achievement is considered a major breakthrough in cold-atom physics.

We conclude this chapter by considering the crossover between two very different extreme regimes for superfluidity in a two-component Fermi gas. The first is the Bardeen-Cooper-Schrieffer (BCS) regime, where the normal state is a degenerate two-component Fermi liquid that undergoes Cooper pairing. The second regime is a BEC of bosons which are composite objects made up of an even number of fermions. Magnetic-field Feshbach resonances make it possible to explore the crossover from a superfluid phase of weakly bound fermions to Bose-Einstein condensation of strongly bound composite bosons.

Chapter 2 presents our results for coherent transport by fermions through arrays of potential wells. We use a Green's function approach to calculate the Density-of-States (DOS) for both single-period and double-period arrays and for a Fibonacci-ordered quasi-periodic array. The practical tool for this calculation is a decimation-renormalization scheme, which we explain in detail. The DOS calculation is then extended to treat an array inserted between two infinite leads. We will see how these leads affect the DOS and the transmission coefficient through the array. We present numerical results for two types of situation: (i) transport by Fermi-surface electrons through arrays of quantum dots; and (ii) transport of atomic gases through optical lattices. The case of fermionic transport is also contrasted with that by a Bose-Einstein condensate.

Chapter 3 is devoted to study the ground state of two-component Fermi gases with repulsive interactions subject to external confining potentials inside 1D optical lattices. We first review some recent work by Gao Xianlong *et al.* [Phys. Rev. B **73**, 165120 (2006)] on unpolarized gases, highlighting in particular phase separation between metallic and Mott-insulating phases. We then present original numerical results for the spin-resolved atom-density profiles obtained from Kohn-Sham spin-density-functional calculations that employ a local-spin-density approximation based on a Bethe-*ansatz* solution. We analyze both spin-independent and spin-dependent parabolic external potentials. In the latter case we find that phase separation of the two spin populations occurs in a certain region of parameters.

Chapter 4 deals with Fermi gases with attractive interactions subject to parabolic trapping inside 1D optical lattices. Motivated by recent experiments, carried out at MIT and at Rice University, on Fermi gas with imbalanced spin populations, we review the theory of various exotic bulk superfluid phases such as the Sarma phase, the Fulde-Ferrell-Larkin-Ovchinnikov phase, and the breached-pair superfluid phase. Recently Gao Xianlong *et al.* [cond-mat/0609346] have analyzed unpolarized fermions with attractive interactions confined by a parabolic potential inside a 1D optical lattice: they have shown that the harmonic confinement induces the emergence of atomic-density waves. In this Thesis we study the same physical system allowing though a finite spin polarization. We observe that a sizeable fraction of majority-spin atoms accumulates at the edges of the trap, leaving a core of paired atoms at the center of the trap. To further investigate the nature of this "phase-separated" ground state, we solve mean-field Bogoliubov-de Gennes equations, which provide information on both the spin-resolved density profiles and the local pairing gap. We observe that the local pairing gap (i) is in the unpolarized case a flat function of position in the bulk of the trap; (ii) oscillates at the edges of the trap as soon as a spin imbalance is introduced, while remaining flat in the bulk of the trap; and (iii) becomes a highly-oscillating function everywhere in the trap with increasing spin-polarization, eventually approaching zero in the fully-spin-polarized case.

Acknowledgments

During my Ph.D. studies at Scuola Normale Superiore di Pisa, I have been influenced professionally and personally by more people that I could possibly list here.

First of all, I would like to thank my advisor Prof. Mario Tosi for giving me the chance to do my Ph.D. with him. He has been the greatest example of a real scientist that has spent his entire life for research. During my Ph.D. he has showed me how to extract physical concepts from the raw calculations and data. Continuously, since I began to write my dissertation, he has showed me how to be precise in organizing and presenting a scientific document. It is a great honour for me to be one of his last students in his prolific career.

I have worked with other members of Prof. Tosi's group in the past 3 years: Marco Polini, Patrizia Vignolo, Pablo Capuzzi, Gao Xianlong and Saeed Abedinpour. The work in Chapter 2 has been done with Patrizia with whom I am indebted for all her patience and tolerance. Gao is a real hard-worker researcher and he has clarified many physical and computational points for me.

Most of my work has been done with Marco Polini with whom I am indebted a lot. As an officemate we have worked together for a long period day by day and he has followed all my improvements continuously. As a physicist I am impressed by his huge effort and also vast knowledge of condensed matter physics. As one of my best friends also, I appreciate his hospitality with his wife Laura Maley and his extremely hospitable parents in Siracusa.

I deeply acknowledge Pablo Capuzzi. He incredibly knows everything about everything in computational physics! As the ultimate debugger, he has been always patient to my endless questions. I wish him the best for his new position in Argentina.

A special thank goes also to Prof. Rosario Fazio and Matteo Rizzi, who have provided us with their DMRG data that I have included in Chapters 3 and 4.

I had a chance also to meet Prof. Klaus Capelle in his short visits to Pisa. Despite the short temporal overlap I had the chance to learn more about DFT.

Living in a new country can be always problematic. After my arrival in September 2003, the former members of the group, Bahman Davoudi and Mehran Asgari helped me a lot to systematize my new life. Bahman, who was my former mentor in Iran, left Pisa shortly after I arrived. With Mehran we spent memorable time until his departure in the summer of 2004. I thank both of them for all their openness and kindness.

Special thank goes to Beatrice Penati: she has been teaching me step-by-step the lovely Italian language as the most fundamental need to join the new society. She has been very close friend over the most difficult moments. Her extraordinary will to explore new frontiers plus her care, kindness and hospitality will remain forever in my mind.

I also would like to thank *some* of my friends that I enjoyed their accompany during these years: Saeed Abedinpour, Sven Winklmann, Riccardo Catena, Janine Splettstoesser, Solmaz Arvani, Masoud Amirkhani, Sohail Sharifi, Shabnam Safaei, Mark Demers, Tatiana Korneeva, Alessandro Poggio, Ilaria Gabbani, Anna Cannavo, Alvaro Garcia Camacho, Alessandro Soru, Tanja Trska, Tristan Hedrick, Gianluigi Del Magno and Anka Ziefer. I thank in particular my long-lasting flatmates Fabio Pacciolla and Paola D'Aleo.

And my *family*: after high school I have been far away from them to study. During all these years they have been an endless resource of support to attain a better future. My parents, the best teachers ever, taught me the most essential lessons of life: simplicity, sincerity and effort. Without them and my lovely siblings, no achievements could have been possible. I dedicate this thesis to all of them.

Chapter 1
Ultracold atomic gases

The main aim of this chapter is to present some basic ideas regarding the physics of atomic quantum gases in low-dimensional configurations. Some of the experimental techniques used for the realization of these gases will be briefly discussed in Section 1.2. Attention is devoted in the final part of the chapter to the topic of quantum phase transitions in cold atoms trapped in optical lattices.

1.1. Introduction to relevant theoretical facts

In this section, some main theoretical aspects concerning low-dimensional Bose and Fermi atomic gases are reviewed. The covered subjects have been taken from the existent exhaustive literature [1, 2, 3].

1.1.1. Bose-Einstein condensation and superfluidity

In 1924 Bose interpreted Planck's law as a statistical distribution function for a photon gas. Immediately afterwards Einstein extended the theory to an ideal gas of integer-spin massive particles by proposing an expression for the average occupation of a quantum single-particle state. From this expression Einstein predicted that below some density-dependent temperature the population of the ground state would reach macroscopic values. This phenomenon is called Bose-Einstein condensation and has been considered for many years as a theoretical model which would be hard to realize in the laboratory.

Bose-Einstein condensation can occur when a gas of bosonic atoms is cooled down to the point where the de Broglie wavelength $l_{dB} = \hbar/mv_T$ becomes comparable with the mean interparticle separation $d \simeq n^{-1/3}$. Here v_T is the atomic thermal speed, m is the atomic mass, and n is the atomic number density. Under these conditions the atomic wave packets overlap and quantum interference between identical particles becomes crucial in determining the statistical behavior of the gas. A phase transition leads to the formation of a Bose-Einstein condensate (BEC), namely to a coherent cloud of atomic matter in which a macroscopic number of atoms occupy the same quantum state, thereby forming a sort of "giant matter wave".

The transition temperature and the peak atomic density in an ideal gas are related as

$$nl_B^3 \sim 2.612. \tag{1.1}$$

At a typical density $n = 10^{12} - 10^{14}$ particles/cm^3, the BEC starts to form in alkali gases at temperatures around 100 nK, which is seven orders of magnitude lower than for the superfluid state in liquid ^{4}He. At such ultra-low temperatures classical interactions would localize the atoms and prevent quantum overlaps, on a time scale which is dictated by binary and higher collisions leading to the formation of clusters and superatomic aggregates. In the experiments yielding a BEC [4, 5], however, the gas is cooled along an out-of-equilibrium path which quenches the kinetic energy of the atoms while they are still in the gaseous state. Three-body and higher collisions are the main processes which limit the lifetime of the condensate to a few seconds, but this is long enough to perform experiments in a metastable state under quasi-equilibrium conditions.

In current experiments the gas is subjected to an external potential due to magnetic or optical fields that are used to cool and hold the atoms. The final output of the cooling process is a mesoscopic phase-coherent droplet of micrometer size. The inhomogeneity due to the external confinement deeply affects the physical properties of the gas, such as its spatial distribution (with a spectacular condensate peak appearing at the trap center), its thermodynamic laws, its excitation spectrum, and its behavior at the phase transition boundary [1].

The introduction of a new length and energy scale by the confinement allows for new equilibrium states that are not available in the homogeneous macroscopic limit and modifies the role of fluctuations.

Such inhomogeneous confinement is usually well approximated by an external potential obeying a harmonic law (superposed in the case of an optical lattice to a sinusoidal law), and is a key ingredient in the characterization of the gas. Current traps are extremely versatile: they can be tuned in space to vary the geometry of the confining potential and periodically varied in time, or suddenly turned off to allow free expansion of the gas. The interplay between external potential and atom-atom interactions gives rise to a variety of new physical effects which are amenable to observation.

The realization of ultracold atomic gases originated as an application of very precise and well controlled laser-beam techniques, thus providing an alternative to the usual buffer-gas technique of cooling by contact with a cold reservoir [6]. The first proposal for "laser-cooling" of free atoms was based on the Doppler effect [7]. An atom moving in a weak standing wave, which is slightly detuned to the red away from an atomic transition, can absorb a photon from the counterpropagating laser wave and re-emit it by spontaneous emission in a random direction (see Section 1.2.1). The atom slows down and the final temperature T_D of the gas, with the meaning of an average kinetic energy, is determined by the natural width of the excited state ($T_D \simeq 140\,\mu K$ for ^{87}Rb). Other laser-cooling mechanisms that lead to temperatures well below the Doppler limit have later been proposed and realized. The so-called Sisyphus cooling [8] uses a laser polarization gradient which removes the degeneracy of the atomic ground-state sublevels as a function of the spatial position: an atom is pumped from a sublevel to another as it moves in space and loses kinetic energy. The temperature limit T_R for such a mechanism is determined by the recoil energy in the emission of a photon ($T_R \simeq 0.3\,\mu K$ for ^{87}Rb). This limit has been overcome for a gas of metastable ^{4}He atoms by using a laser-cooling scheme based on a velocity-selective optical pumping of atoms into a non-absorbing coherent superposition of states [9]. However, such a scheme is impracticable for atoms with a richer internal structure [10] and cannot be applied to high-density spatial distributions since collisions drive the atoms out of the non-absorbing state [11].

Laser cooling of the gaseous cloud is followed by evaporative cooling inside a magnetic trap [12]. The method consists of progressively eliminating the "hot" atoms from the trap while allowing thermalization of the remaining atoms *via* elastic collisions. The cooling process competes with heating due to losses, which are mainly in the three-body channel close to the condensation region, at relatively high densities. The condition for Bose-Einstein condensation, as expressed in terms of the phase-space density $n\,l_{dB}^3$ is extremely severe: for alkali atoms it requires reaching temperatures below $T \sim 0.1\,\mu K$ and densities $n \sim 10^{13}$ atoms/cm^3.

After the evaporative cooling cycle the condensate appears as a high-density peak at the center of the density distribution of the atomic cloud, which can be imaged by a resonant absorption technique after release from the trap and ballistic expansion (see Figure 1.1). Due to the inhomogeneity introduced by the confinement it is thus possible to observe "condensation" also in real space: roughly speaking, the condensate populates macroscopically the lowest level of the harmonic trap, whose wavefunction is localized around the trap center. A detailed account of the route to BEC can be found in the 2001 Nobel Lectures [13, 14].

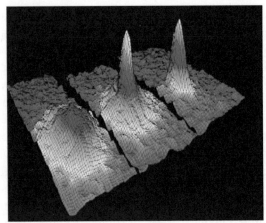

Figure 1.1. The first BEC in a gas of ^{87}Rb atoms, as it appears in two-dimensional time-of-flight images as temperature is decreased below T_c. From Anderson *et al.* [4].

Low-dimensional Bose gases

Turning to the role of trap dimensionality, the subject of low-dimensional quantum gases has a long history. The influence of dimensionality of the system of bosons on the presence and character of Bose-Einstein condensation and superfluid phase transition has been a subject of extensive studies in the spatially homogeneous case. From a general point of view, BEC is impossible in one-dimension (1D) and (except at $T = 0$) in two-dimension (2D) in a homogeneous (macroscopic, infinitely extended) system, but should occur in atom traps because of the changes in the density of states that finite system sizes induce [15].

The earlier discussiones of low-dimensional Bose gases were mostly academic as there was no possible realization of such systems. Fast progress in evaporative and optical cooling of trapped atoms and the observation of BEC of alkali atoms in trapped clouds stimulated a search for non-trivial trapping geometries. Present facilities allow one to tightly confine the motion of trapped particles to zero point oscillations in one or two directions. Then, kinetically the gas is 2D or 1D, and the value of the effective interparticle interaction depends on the tight confinement.

Recent experiments have already reached 2D and 1D regimes for trapped Bose gases and studied some of the quantum degenerate states. These studies bring to light new physics originating from finite size, spatial inhomogeneity, and finite temperature. The present section covers some important issues in the physics of 2D and 1D trapped quantum gases: the nature of various quantum degenerate states, the role of the interactions between the particles, and the role of finite-temperature and finite-size effects.

We start by describing a crossover to the BEC regime in ideal 2D and 1D Bose gases with a finite number of particles. We will consider an equilibrium gas at temperature T in the grand-canonical ensemble, where the chemical potential μ is fixed and the number of particles N is fluctuating. In the thermodynamic limit ($N \to \infty$) this is equivalent to the description in the canonical ensemble (fixed N).

In any dimension and confining potential, the ideal Bose gas is characterized by a set of eigenenergies for a single particle, E_ν, with the index ν labeling quantum numbers of the particle eigenstates. The (average) total number of particles N is then related to the temperature and chemical potential by the equation

$$
\begin{aligned}
N &= \sum_\nu N_\nu \\
&= \sum_\nu \frac{1}{e^{\beta(E_\nu - \mu)} - 1},
\end{aligned}
\tag{1.2}
$$

where N_ν are the equilibrium occupation numbers of the eigenstate and β is related to T and Boltzman's constant k_B as $\beta = (k_B T)^{-1}$. We now see how equation (1.2) allows one to establish the presence or absence of BEC in 2D and 1D ideal Bose gases. We consider an ideal Bose gas both in the homogeneous case and confined in a harmonic trap.

Homogeneous ideal gas

In an infinitely extended *uniform* gas, the particle eigenstates are characterized by the momentum $\mathbf{k}$ and the energy $E_k = \hbar^2 k^2/(2m)$ where m is the mass of a particle. Then equation (1.2) takes the form

$$
N = \Omega \int \frac{d^d k}{(2\pi)^d} \frac{1}{e^{\beta(E_k - \mu)} - 1},
\tag{1.3}
$$

with d being the dimension of the system, and Ω the d-dimensional volume

In the 2D case the integration in equation (1.3) is straightforward and we obtain

$$
\mu = k_B T \ln\left[1 - \exp(-n_2 \Lambda_T^2)\right] < 0,
\tag{1.4}
$$

where $\Lambda_T = (2\pi\hbar^2/mT)^{1/2}$ is the thermal de Broglie length and n_2 is the 2D density. The quantity $n_2 \Lambda_T^2$ is called the degeneracy parameter and in 2D it can be written as

$$
n_2 \Lambda_T^2 = \frac{T}{T_d},
\tag{1.5}
$$

where

$$T_d = \frac{2\pi\hbar^2}{m}n_2 \tag{1.6}$$

is the temperature of quantum degeneracy. In the limit of a classical gas, equation (1.4) gives the well-known result

$$\mu = k_\mathrm{B}T \ln(n_2\Lambda_T^2). \tag{1.7}$$

For a strongly degenerate gas, where $n_2\Lambda_T \gg 1$, we obtain

$$\mu = -k_\mathrm{B}T \exp(-n_2\Lambda_T^2). \tag{1.8}$$

Unlike the 3D case, the function $\mu(T)$ is analytical and shows a monotonic increase of the chemical potential with decreasing temperature down to $T \to 0$. In the thermodynamic limit the population of the ground state ($k = 0$) remains microscopic. One thus can say that there is no BEC at finite temperature in an ideal uniform 2D Bose gas.

The situation is similar for an infinite uniform 1D Bose gas, where the degeneracy parameter is $n_1\Lambda_T = (T/T_d)^{1/2}$ and the temperature of quantum degeneracy is given by

$$T_d = \frac{2\pi\hbar^2}{m}n_1, \tag{1.9}$$

with n_1 being the 1D density. In the classical limit ($n_1\Lambda_T \ll 1$) and in the limit of a strongly degenerate gas ($n_1\Lambda_T \gg 1$), equation (1.3) gives

$$\mu = k_\mathrm{B}T \ln(n_1\Lambda_T), \qquad n_1\Lambda_T \ll 1 \tag{1.10}$$

$$\mu = -\frac{\pi k_\mathrm{B}T}{(n_1\Lambda_T)^2}. \qquad n_1\Lambda_T \gg 1 \tag{1.11}$$

Again, the chemical potential monotonically decreases with temperature and remains negative at any T, which indicates the absence of BEC.

The absence of BEC in macroscopic (uniform) 2D and 1D Bose gases presents a striking difference from the 3D case. This difference originates from the energy dependence of the density of states. The density of states is

$$\rho(E) \propto E^{(d/2-1)}, \tag{1.12}$$

and for a 3D gas it increases with E. Therefore, at sufficiently low temperature it becomes impossible to thermally occupy the low-energy states while maintaining a constant chemical potential or density. As a result, a macroscopic number of particles goes into the ground state ($k = 0$), i.e. one has the phenomenon of BEC. In 2D and 1D the density of states does not increase with E and the phenomenon of BEC is absent.

Ideal gas in a harmonic trap

For 2D and 1D ideal Bose gases in a harmonic confining potential, the density of states is $\rho(E) \propto E^{(d-1)}$ and the situation changes. The population of the ground state $(E = 0)$ is

$$N_0 = \frac{1}{\exp(-\beta\mu) - 1}. \tag{1.13}$$

At finite T, for a large but finite number of particles in a trap, N_0 can become macroscopic at a small but finite negative μ. One then speaks of a crossover to the BEC regime.

We first discuss the BEC crossover for the 2D Bose gas in a symmetric harmonic confining potential $V(\mathbf{r}) = m\omega^2(x^2 + y^2)/2$. In this case the particle energy is $E_\nu = \hbar\omega(n_x + n_y)$, with quantum numbers n_x, n_y being non-negative integers, and ω is the trap frequency. The density of states is then $\rho(E) = E/(\hbar\omega)^2$ therefore, separating out the population of the ground state, one can replace the summation in equation (1.2) by an integration:

$$N = N_0 + \int_0^\infty dE\rho(E)\frac{1}{e^{\beta(E-\mu)} - 1} \tag{1.14}$$

Assuming a large population of the ground state, from equation (1.13) we obtain $-\mu/(k_B T) \approx 1/N_0 \ll 1$, and the population of excited trap states proves to be

$$\int_0^\infty dE\rho(E)\frac{1}{e^{\beta(E-\mu)} - 1} \approx \left(\frac{k_B T}{\hbar\omega}\right)^2\left(\frac{\pi^2}{6} - \frac{1 + \ln N_0}{N_0}\right). \tag{1.15}$$

This allows one to write equation (1.14) in the form

$$N\left[1 - \left(\frac{T}{T_c}\right)^2\right] = N_0 - \left(\frac{k_B T}{\hbar\omega}\right)^2\left(\frac{1 + \ln N_0}{N_0}\right) \tag{1.16}$$

where

$$T_c = \sqrt{\frac{6N}{\pi^2}}\hbar\omega \tag{1.17}$$

For a large number of particles, equation (1.16) indicates the presence of a sharp crossover to the BEC regime at $T \approx T_c$. Below T_c we omit the last term in equation (1.16) and obtain the occupation of the ground state

$$N_0 \approx N\left[1 - \left(\frac{T}{T_c}\right)\right] \tag{1.18}$$

The result of equation (1.18) is similar to that in the 3D case and was first obtained by Bagnato and Kleppner [15].

Above T_c the first term on the right hand side (rhs) of equation (1.16) is zero. The width of the cross-over region, ΔT, *i.e.* the temperature interval where both terms on the rhs of equation (1.16) are equally important, is given by

$$\frac{\Delta T}{T_c} \sim \sqrt{\frac{\ln N}{N}} \tag{1.19}$$

The condensed fraction of particles, $N_0(T)/N$, following from equation (1.16), is presented in Figure (1.3) for various values of N. For a large N the crossover region is very narrow and one can speak of an ordinary BEC transition in an ideal harmonically trapped 2D gas.

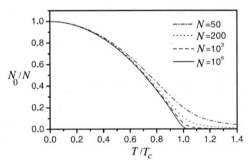

Figure 1.2. The ground state population in a 2D trap versus temperature, calculated from equation (1.16). Figure adapted from [2].

For 1D ideal Bose gas in a harmonic potential $V(z) = \frac{1}{2}m\omega^2 z^2$, the particle energy is $E = j\hbar\omega$, with j being a non-negative integer and the density of states is $\rho(E) = 1/\hbar\omega$. Here, the integral representation of equation (1.14) fails as the integral diverges for $\mu \to 0$. Therefore, we should correctly take into account the discrete structure of the lowest energy levels. In the limit $\{-\mu, \hbar\omega\} \ll k_B T$ we rewrite equation (1.2) in the form

$$
\begin{aligned}
N = N_0 &+ \frac{k_B T}{\hbar\omega} \sum_{j=1}^{M} \frac{1}{j - \mu/\hbar\omega} \\
&+ \sum_{j=M+1}^{\infty} \frac{1}{\exp(\hbar\omega j/k_B T - \mu/k_B T) - 1}
\end{aligned} \tag{1.20}
$$

where the number M satisfies the inequalities $1 \ll M \ll k_B T/\hbar\omega$. The first term is

$$\sum_{j=1}^{M} \frac{1}{j - \mu/\hbar\omega} = \psi(M + 1 - \mu/\hbar\omega) - \psi(1 - \mu/\hbar\omega)$$

$$\approx -\psi(1 - \mu/\hbar\omega) + \ln(M - \mu/\hbar\omega),$$

(1.21)

where ψ is the digamma function. This function is defined as [16]:

$$\sum_{k=1}^{n} \frac{1}{k - a} = \psi(n + 1 - a) - \psi(1 - a),$$

(1.22)

where

$$\psi(n) = -\gamma_0 + \sum_{k=1}^{n-1} \frac{1}{k},$$

(1.23)

and $\gamma_0 \simeq 0.572$ is the Euler constant. The second term in equation (1.20) can be transformed to an integral ,

$$\sum_{j=M+1}^{\infty} \frac{1}{\exp(\hbar\omega j/k_B T - \mu/k_B T) - 1}$$

(1.24)

$$\approx \frac{k_B T}{\hbar\omega} \int_{\hbar\omega M/k_B T}^{\infty} \frac{dx}{\exp(x - \mu/k_B T) - 1}$$

$$\approx -\frac{k_B T}{\hbar\omega} \ln \frac{\hbar\omega(M - \mu/\hbar\omega)}{k_B T}$$

(1.25)

Finally, since in the limit of $|\mu| \ll k_B T$ the chemical potential is related to the population of the ground state as $-\mu \approx k_B T/N_0$, we reduce equation (1.20) to the form

$$N - \frac{k_B T}{\hbar\omega} \ln\left(\frac{k_B T}{\hbar\omega}\right) = N_0 - \frac{k_B T}{\hbar\omega} \psi\left(1 + \frac{k_B T}{\hbar\omega N_0}\right)$$

(1.26)

As in 2D case, we have two regimes, with the border between them at a temperature

$$k_B T_{1D} \approx \frac{N}{\ln N} \hbar\omega.$$

(1.27)

For the temperature below T_{1D}, the first term on the rhs of equation (1.26) greatly exceeds the second one and the ground-state population behaves as

$$N_0 \approx N - \left(\frac{k_B T}{\hbar\omega}\right) \ln\left(\frac{k_B T}{\hbar\omega}\right).$$

(1.28)

The crossover region is determined as the temperature interval where both terms are equally important:

$$\frac{\Delta T}{T_{1D}} \sim \frac{1}{\ln N}. \tag{1.29}$$

In contrast to the 3D and 2D cases, the crossover temperature is much lower than the degeneracy temperature $k_B T_d \approx N\hbar\omega$. These results have been obtained by Ketterle and van Druden [17]. Figure 1.3 presents the relative occupation of the ground- state $N_0(T)/N$ calculated from equation (1.26).

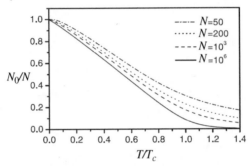

Figure 1.3. The ground state population in a 1D trap versus temperature, from equation (1.26). Figure adapted from [2].

Interacting Bose gas

Spatially uniform 1D Bose gases with repulsive interparticle interaction have been extensively studied in the last decades. For the delta-functional interaction, Lieb and Liniger [18] have calculated the ground state energy and the excitation spectrum. Generalizing the Lieb-Liniger approach, Yang and Yang [19] have proved analyticity of thermodynamic functions at any finite temperature T, which indicates the absence of a phase transition. Here we focus qualitatively on the strongly interacting regime for the 1D system and its experimental realization.

The physics of ultracold 1D Bose systems is very different from that of ordinary 3D cold gases. By decreasing the particle density n, a 3D quantum many-body system becomes more ideal, whereas in a 1D Bose gas the role of interactions becomes more important. The reason is that for a 1D Bose gas at temperatures $T \to 0$, the kinetic energy of a particle is $K \approx \hbar^2 n_1^2/m$ and the interaction energy per particle is $I = g n_1$, where g is the 1D delta-function repulsive interaction strength,

$$V(x_i, x_j) = g\, \delta(x_i - x_j). \tag{1.30}$$

We see that the kinetic energy decreases with decreasing density faster than the interaction energy. The ratio of the interaction to kinetic energy,

$$\gamma = \frac{mg}{\hbar^2 n_1}, \tag{1.31}$$

characterizes the different physical regimes of the 1D quantum gas.

In the weakly interacting regime, the wavefunction of particles is not influenced by the interaction and one should have $I \ll K$. This leads to the criterion of the weakly interacting regime as

$$\gamma = \frac{mg}{\hbar^2 n_1} \ll 1 \tag{1.32}$$

From equation (1.32) one really sees that in contrast to 3D gases, the 1D Bose gas becomes more interacting with decreasing density. A qualitative illustration of different interactive regimes is shown in Figure 1.4.

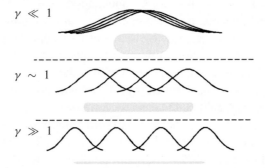

Figure 1.4. Cartoon of 1D boson distributions for three different γ regimes. As γ increases, the bosons, like fermions, become spatially distinct. The three shaded drawings represent the atomic density in a 1D tube. The Tonks-Girardeau regime corresponds to $\gamma \gg 1$. Figure adapted from [20].

At large values of γ the gas enters the Tonks-Girardeau (TG) regime, where the repulsion between particles strongly decreases the wave-function at short interparticle distances. The model of the TG gas proposed about 40 years ago [18, 21]. In such a gas, the repulsive interactions between bosonic particles confined to 1D dominate the physics of the system. In order to minimize their mutual repulsion, the bosons are prevented from occupying the same position in space. This mimics the Pauli exclusion principle for fermions, causing the bosonic particles to exhibit fermionic properties [18, 21]. However, such bosons do not exhibit completely ideal fermionic (or bosonic) quantum behaviour for example, this is reflected in their momentum distribution [22]. Here we explain the correspondence between the strongly interacting Bose gas and the non-interacting Fermi gas in 1D.

For $\gamma \to \infty$, the ground state of N bosons at zero temperature is described by the many-body wavefunction:

$$\Psi_0(x_1, x_2, \ldots, x_N) \propto |\det[\varphi_i(x_j)]| \qquad (1.33)$$

where det denotes the Slater determinant, and $\varphi_i(x_j)$ is the i-th eigenfunction of the single-particle hamiltonian. Of course, the Slater determinant is the wave function of the ideal spin-polarized (or "spinless") fermion gas. The form of the Slater determinant guarantees that the wavefunction vanishes whenever two particles occupy the same position in space. However, the absolute value of the determinant ensures that the wavefunction for the bosons remains completely symmetric. However, some properties are strongly modified by the presence of the absolute value of the determinant, leading to a unique behaviour of the momentum distribution of the TG gas. This can be understood qualitatively in the following way: the bosonic particles in a TG gas are not allowed to occupy the same position in space. Owing to this restriction, they are distributed over a more extended region in momentum space than in the case of an ideal or weakly interacting Bose gas. On the other hand, in order to keep themselves apart from each other, they do not need to be in different momentum states, as would be the case for fermions.

The TG regime has been achieved experimentally in an optical lattice [23]. In this experiment, a novel way to achieve the TG regime has been proposed and demonstrated . The main point is to include an additional optical lattice along the 1D gas, which results in an increase of γ which for a homogeneous gas is expressed by equation (1.31). Larger values of γ could either be reached by decreasing the density of the quantum gas or by increasing the effective interaction strength between the particles. The addition of a periodic potential along the third axis increases the effective mass, and thus leads to an increase of γ. Following these ideas, these authors have been able to enter the TG regime with $\gamma = 5 - 200$. In this regime, the bosons can be theoretically described using a 'fermionization' approach. In the experiment the bosonic atoms exhibit a pronounced fermionic behaviour, and show a momentum distribution that is in excellent agreement with a theory of fermionized trapped Bose gases [23].

In another experiment, a quasi-1D Bose gas has been demonstrated in highly elongated traps with an aspect ratio of order 100 [24]. The 1D regime is reached when the condition

$$\{\mu, k_B T\} \ll \hbar\omega \qquad (1.34)$$

is fulfilled , where ω denotes the strength of the radial harmonic confinement. In this experiment, up to 2×10^9 ^{87}Rb atoms has been collected in a magneto-optical trap and the gas is characterized by measuring its lowest-lying collective excitation [24].

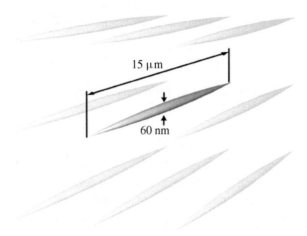

Figure 1.5. The geometry and size of trapped 1D gases. From [24].

1.1.2. Key features of trapped atomic Fermi gases

Fermions, such as electrons, protons, and neutrons, compose all of the matter around us. Unlike bosons, they have half integer spin and constitute the second half of the particle family tree. To tell whether an atom is a boson or a fermion, we look at the total number of protons, neutrons, and electrons making up that atom. Since these are all spin-$\frac{1}{2}$ fermions, adding up an odd number of them will make an atom that is a fermion (half-integer spin).

The Pauli exclusion principle, which forbids identical fermions from occupying the same quantum state, must be satisfied for fermions. This property gives rise to remarkable effects that include the structure of the periodic table of the elements, the nature of electrical conductivity in metals, and the quantum Hall effect. White-dwarf and neutron stars manifest the Pauli principle in a striking way. In both cases, the star has exhausted its nuclear fuel and is essentially dead. Tremendous gravitational forces draw the star in on itself. Inside such a star, however, there is near-unity occupation of the available quantum states (*i.e.* quantum degeneracy), which provides a *"Fermi pressure"* that stabilizes the star against collapse.

In a gas of N identical fermions the occupation probability $f(\varepsilon)$ for a single-particle state with energy ε is given by the Fermi-Dirac (FD) distribution

$$f(\varepsilon) = \frac{1}{\exp\left[(\varepsilon - \mu)/k_{\mathrm{B}}T\right] + 1}. \tag{1.35}$$

where the chemical potential μ is fixed by the normalization condition

$$N = \int g(\varepsilon)f(\varepsilon)d\varepsilon. \tag{1.36}$$

The density of states, $g(\varepsilon)$, e.g. for a three-dimensional confining potential $V(r) = \frac{1}{2}m\omega^2 r^2$ is given by

$$g(\varepsilon) = \frac{\varepsilon^2}{2(\hbar\omega)^3}. \tag{1.37}$$

At zero temperature, $T = 0$, the ensemble of fermions forms a so-called Fermi sea where each quantum state is filled up to the Fermi energy

$$E_{\mathrm{F}} = \mu. \tag{1.38}$$

The corresponding temperature to this energy is referred to as the Fermi temperature

$$T_{\mathrm{F}} = \frac{E_{\mathrm{F}}}{k_{\mathrm{B}}}. \tag{1.39}$$

The Fermi energy is obtained by integrating equation (1.36) at $T = 0$. In case of fermions with opposite spin-$\frac{1}{2}$ it gives

$$E_{\mathrm{F}} = \hbar\omega \left(6N\right)^{1/3}. \tag{1.40}$$

As an example, 10^5 identical fermions harmonically trapped with a trapping frequency $\omega = 2\pi \times 100$ Hz correspond to a Fermi energy about 400 nK. It is also useful to define a Fermi wavenumber $k_{\mathrm{F}} = \sqrt{2mE_{\mathrm{F}}}/\hbar$. This wavenumber can be expressed in term of fermions number N [25],

$$k_{\mathrm{F}} = \frac{1}{\hbar}\left[2m\,\hbar\omega(6N)^{1/3}\right]^{1/2}. \tag{1.41}$$

The fermionic ensemble in a trap and its difference from bosonic one are shown in Figure 1.6

Another difference between bosons and fermions is that fermions do not undergo a sudden phase transition in the ultra-low temperature regime. Instead, the quantum behavior emerges gradually as the fermion gas is cooled below the Fermi temperature T_{F}. This temperature which is typically less than 1 μK for atomic gases, marks the crossover from the classical to the quantum regime.

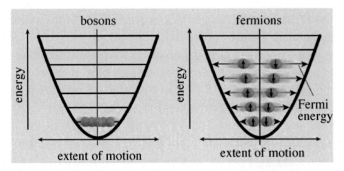

Figure 1.6. All atoms in a Bose-Einstein condensate attain the same energy level, the lowest one available (left), whereas only two atoms (with opposite $\frac{1}{2}$ spin) can share one energy level in a degenerate Fermi gas (right). As a result, the atoms in such a gas occupy a series of increasing energy levels, up to a level that corresponds to the Fermi temperature of the gas. The higher the energy level of an atom, the broader its oscillatory motion within the trap. From [26].

Cooling fermionic atoms

The main experimental challenge in creating a fermionic gas is to chill the atoms to ultra-low temperatures. Given the fact that physicists have been able to cool bosons to microkelvin temperatures since 1995, when BEC was first created, one might think that experiments to cool fermions would have followed soon afterwards. However, cooling a gas of fermionic atoms is difficult because of their collision properties. Elastic collisions keep the gas in thermal equilibrium as it cools, and also affect many properties of both Bose and Fermi gases by the effects of the interparticle interactions. However, differences in the collisional behavior of bosonic and fermionic atoms arise at temperatures well above those that are required for the gas as a whole to behave quantum mechanically.

The main type of collisions between atoms in an ultracold gas is from binary s-wave scattering events where there is no relative angular momentum between the two atoms. But due to the Pauli exclusion principle, these s-wave collisions are forbidden for fermions that are in the same internal quantum state (in brief, the same spin state). This lack of collisions makes it impossible to cool such a fermionic atomic gas efficiently.

However, fermionic atoms that are in *different* internal states can undergo s-wave collisions: the space part of the wave function is in this case symmetric and the overall wave function, written as the product of spin and space parts, is antisymmetric. Using a two-component Fermi gas (or a boson-fermion mixture) it is then possible to cool fermions to near absolute zero. The first quantum-degenerate Fermi gas of atoms was created by DeMarco and Jin in 1999 at JILA [27]. In the following we point out the main features of this achievement.

In the first experimental realization, an evaporative cooling strategy that uses a two-component Fermi gas was employed to cool a magnetically trapped gas of 7×10^5 ^{40}K to 0.5 of the Fermi temperature T_F. DeMarco and Jin trapped a mixture of atoms in two magnetic sublevels, $|F = 9/2, m_F = 9/2\rangle$ and $|F = 9/2, m_F = 7/2\rangle$, of the hyperfine ground state having total atomic spin $F = 9/2$ (m_F is the magnetic quantum number). By keeping the mixture stably confined, DeMarco and Jin evaporatively cooled it to below 300 nK and obtained information about quantum degeneracy by turning the magnetic trap off, allowing the gas to expand and measuring the shadow of the gas cast by a laser. These absorption images determine the energy, temperature, number and momentum distribution of the atoms in the gas.

This approach, pioneered by Jin's group at JILA, is not the only way of obtaining ultracold Fermi gases. Some other group [28, 29] have used a technique known as "sympathetic cooling", in which a gas containing a mixture of isotopes - rather than a mixture of spin-states - is cooled. Both groups have used this approach to cool mixtures of ^{6}Li and ^{7}Li atoms. The ^{7}Li atoms, which are bosons, are evaporatively cooled in the usual way. Meanwhile, the simultaneously trapped ^{6}Li atoms, which are fermions, cool simply by being in thermal contact with the boson gas. Using these novel experimental techniques, physicists have been able to cool Fermi gases of atoms into the quantum-degenerate regime below the Fermi temperature T_F. But because the effects of quantum statistics become stronger as a gas is cooled further into this regime, an ultimate cooling limit is met (due to Pauli blocking: see below).

Quantum degeneracy

The quantum behavior of an atomic Fermi gas was first revealed in thermodynamic measurements. At JILA the atomic Fermi gas was studied by analysing its absorption images after expansion. As the gas expands, fast atoms travel further from the center than slow-moving atoms. An optical image of the gas therefore reveals the momentum distribution of the atoms: those atoms with low momentum remain near the center of the cloud, while atoms with high momentum are found at the edges.

DeMarco and Jin observed [27] the emerging quantum degeneracy in measurements of the total energy and the momentum distribution of the trapped Fermi gas. Classically, at high T, the gas has total energy $U_{cl} = 3Nk_B T$ and a Gaussian momentum distribution. At $T = 0$, however, the atoms occupy the energy levels of the harmonic confining potential in a Fermi sea arrangement with $U_{\mathrm{FD}} = \frac{3}{4}Nk_{\mathrm{B}}T_{\mathrm{F}}$, and the Fermi pressure results in a parabolic momentum distribution [25]. They measured the extra energy due to FD statistics and observed the transition between these two distributions by analyzing the optical depth images of expanded clouds.

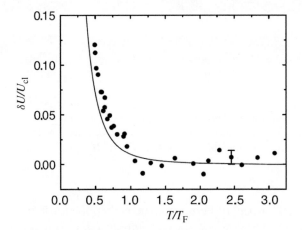

Figure 1.7. Emergence of quantum degeneracy as seen in the energy of the trapped Fermi gas. A moment analysis was used to extract the energy of the gas from time-of-fight absorption images. The excess energy $\delta U = U - U_{\mathrm{cl}}$ is shown versus T/T_{F}, where U is the measured energy and $U_{\mathrm{cl}} = 3Nk_{\mathrm{B}}T$ is the energy of a classical gas at the same temperature. Adapted from DeMarco and Jin [27].

A deviation from classical thermodynamics is exposed in a measurement of the total energy U of the trapped gas. The difference $\delta U = U - U_{\mathrm{cl}}$ between the measured energy and the classical energy at the same T is plotted in Figure 1.7. The temperature T is determined from a fit to the periphery of the absorption image where the effects of the quantum statistics are reduced because of the low mean occupancy at these high-momentum states.

Similar experiments have been carried out to show the differences in size between a Fermi gas (^{6}Li) and a Bose gas (^{7}Li) [30, 31]. In the quantum regime, the mean energy per fermion rises above the value expected from classical physics or in a Bose gas. The fermionic atoms have more kinetic energy, which means that the trapped Fermi gas spreads over a larger volume than the Bose gas (Figure 1.8). This quantum phenomenon, called *Fermi pressure*, is believed to operate in astrophysics and to be responsible for stabilizing white-dwarf and neutron stars against collapsed induced by their gravitational potential.

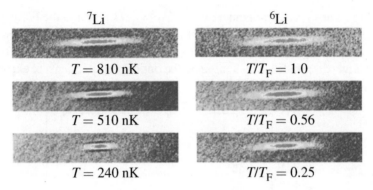

Figure 1.8. Absorption images of a gas of ^{7}Li atoms (bosons) and a gas of ^{6}Li atoms (fermions). The fermions have a higher energy per particle, which means that they move faster and further than the bosons. The Fermi gas therefore occupies a larger volume than the Bose gas as both are cooled in a magnetic trap. From [31].

Another main aspect of Fermi statistic is the relaxion behavior of the oscillations of a fermion gas. In a fermion mixture it is possible to study the collective modes of two spin components [32]. Elementary excitations play a key role in understanding the behavior of quantum fluids. The nature of these excitations varies drastically between two regimes, collisionless and hydrodynamic, based on the relative strength of interactions. In the collisionless regime, the collision rate Γ_{coll} is much smaller than the excitation frequency ω. In this limit, there are few scattering events per oscillation. Classically, the motion is described by the single-particle Hamiltonian and collisions tend to damp excitations. The opposite limit of large collision rate, $\Gamma_{coll} \gg \omega$, is called the hydrodynamic or collisional regime. Here, the motion consists of collective excitations in which the high collision rate maintains local equilibrium throughout the gas [33].

For ^{40}K atoms in a cigar-shaped trap both collisionless and the hydrodynamic regime have been reached in the excitation of dipolar modes, the collisionality parameter in the gas being varied by changing the total number of particles and the radial confinement strength. Experiments performed on the two-component fermion clouds [33, 34] have shown how the frequencies of the two spin dipole modes tend to lock at the same intermediate values as the collisionality goes through a minimum as the gas is driven from the collisionless to the hydrodynamic regime (see Figure 1.9). The quantum fluid displays well-defined collective modes in both regimes, the damping being strong in the intermediate region of collisional rates [1].

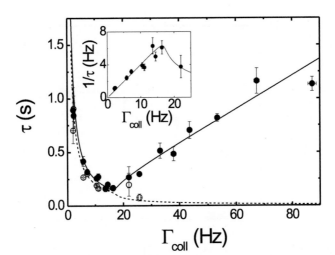

Figure 1.9. Exponential damping times for the two modes of excitation. The damping times exhibit a minimum at the transition from the collisionless to the hydrodynamic regime. In the hydrodynamic regime, the measured τ is seen to depend linearly on Γ_{coll}. The inset shows the damping rate $1/\tau$ which scales linearly with Γ_{coll} in the collisionless regime. Adapted from Gensemer and Jin [33].

The quantum nature of the gas is revealed also through changes in the excitation dynamics as the gas is cooled below T_F. The damping time of the dipole oscillation is measured for an equal mixture of two-component fermion gas of ^{40}K as the temperature of the gas is varied through forced evaporative cooling. The emergence of quantum behavior below T_F has been observed [34] by comparing the measured damping time τ to the classical prediction τ_{class} in the hydrodynamic regime. The measured spin excitation damping time is shown in Figure 1.10. At low T/T_F DeMarco and Jin observed [34] that the damping time decreases significantly compared to the classical expectation.

This change in the damping arise from a quantum statistical suppression of collisions. In the degenerate Fermi gas low energy quantum states have an occupancy approaching one. This fact combined with the Pauli exclusion principle suppresses the elastic collision rate through a reduction of allowed final states. As seen in Figure 1.9, lowering Γ_{coll} in the hydrodynamic regime results in a shorter damping time (or equivalently a higher damping rate). Thus by reducing Γ_{coll}, Pauli blocking [35] hinders the ability of the collective excitation to propagate in the quantum regime. Ultimately, in the zero temperature limit one expects the gas to reach the collisionless regime [34].

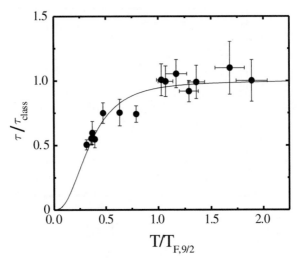

Figure 1.10. Effect of quantum degeneracy on the spin excitation damping time. At low T/T_F the measured damping time τ is reduced compared to the classical expectation τ_{class}. The data agree well with the theoretical prediction from a quantum kinetic calculation (solid line) of the effect of Pauli blocking on the collision rate. Adapted from DeMarco and Jin [34].

1.1.3. Confinement induced resonance in quasi-1D quantum gases

Atom waveguides are a fundamental component of atom optics and are expected to play an important role in atom interferometry. To ensure proper coherence as atomic beams propagate along the waveguides, efforts should be made to avoid decoherence-inducing mechanisms like collisional losses. This requires a detailed understanding of the effects of quasi-1D confinement on atom-atom collision. Such 1D interacting atomic quantum gases have become accessible *via* adiabatic transfer from atomic Bose condensates to highly elongated tight cigar-shape traps. Here the question of an effective 1D coupling constant becomes important. It has been shown [36] that atom-atom scattering under transverse confinement is subject to a confinement-induced resonance (CIR) where the effective 1D coupling strength diverges at a particular ratio of the confinement and scattering length.

The description of the binary collision between cold atoms confined in a waveguide has been proposed [37] *via* the following model:

(a) The waveguide potential is replaced by an axially symmetric 2D harmonic potential of frequency ω_0. The forces created by the potential act in the $X - Y$ plane.

(b) Atomic motion along the Z axis is free.

(c) The interaction between pairs of atoms is modeled by the Huang pseudopotential

$$U(r) = g\,\delta(\mathbf{r})\frac{\partial}{\partial r}(r\ ), \qquad (1.42)$$

where $g = 2\pi\hbar^2 a/m_r$ is the coupling strength, $m_r = m/2$ is the reduced mass, and a is the scattering length. The role of the regularization operator $\frac{\partial}{\partial r}(r\ )$ is in removing the $1/r$ divergence from the scattered wave.

The harmonic nature of the confining potential allows the separation of the center of mass and relative motions. The Schrodinger equation governing the relative motion reads as

$$\left[\frac{\hat{p}_z^2}{2m_r} + g\delta(\mathbf{r})\frac{\partial}{\partial r}(r\ ) + \hat{\mathcal{H}}_\perp(\hat{p}_x, \hat{p}_y, x, y)\right]\Psi = E\Psi, \qquad (1.43)$$

where $\mathbf{r} = \mathbf{r}_2 - \mathbf{r}_1$ is a relative coordinate for atom 1 and 2 and

$$\hat{\mathcal{H}}_\perp = \frac{\hat{p}_x^2 + \hat{p}_y^2}{2m_r} + \frac{1}{2}m_r\,\omega_0^2(x^2 + y^2), \qquad (1.44)$$

is the 2D harmonic oscillator Hamiltonian.

The wave function of the relative coordinate $\mathbf{r} = z\hat{z} + \rho\hat{\rho}$ can be factorized as

$$\Psi(\mathbf{r}) = \phi_0(\rho)\psi(z), \qquad (1.45)$$

where

$$\phi_0(\rho) = \frac{1}{a_\perp\sqrt{\pi}}\exp\left(-\frac{\rho^2}{2a_\perp^2}\right) \qquad (1.46)$$

with $\rho = \sqrt{x^2 + y^2}$. Here

$$a_\perp = \sqrt{\frac{\hbar}{m_r\,\omega_0}}, \qquad (1.47)$$

denotes the transverse harmonic oscillator length, corresponding to the transverse width of the waveguide. The energy spectrum of the 2D harmonic oscillator is

$$E_{n,m_z} = \hbar\omega_0(n + 1), \qquad (1.48)$$

where $n = 0, 1, 2, \ldots, \infty$ is the principal quantum number and the angular momentum with respect to the Z axis, m_z, depends on n as follows:

$$m_z = \begin{cases} 0, 2, 4, \ldots, n & \text{if n is even,} \\ 1, 3, 5, \ldots, n & \text{if n is odd.} \end{cases} \qquad (1.49)$$

The asymptotic form of the scattering wave function Ψ then reads [36]

$$\Psi(z, \rho) \xrightarrow{|z| \to \infty} \left\{ e^{ik_z z} + f_{\text{even}} e^{ik_z |z|} \right\} \phi_0(\rho). \tag{1.50}$$

$f_{\text{even}}(k_z)$ is the 1D scattering amplitude for the even partial waves that should be determined through this approach. After lengthy calculations, the final expression for the 1D scattering amplitude can be written as

$$f_{\text{even}}(k_z) \simeq -\frac{1}{1 - i\,k_z\,a_{1D}}. \tag{1.51}$$

Here

$$a_{1D} = -\frac{a_\perp^2}{2a} \left(1 - \mathcal{C}\frac{a}{a_\perp} \right), \tag{1.52}$$

is the one-dimensional scattering length and $\mathcal{C} \simeq 1.4603$ [36].

It is now tempting to introduce an effective 1D interaction in such a way that its scattering amplitude matches equation (1.51), *i.e.* to solve the corresponding 1D inverse scattering problem [38]. Such an object does exist and is represented by a 1D δ-function potential [36]

$$U_{1D}(z) = g_{1D}\,\delta(z), \tag{1.53}$$

with a coupling strength

$$g_{1D} = -\frac{\hbar^2}{m_r a_{1D}} = \frac{g}{\pi a_\perp^2} \left(1 - \mathcal{C}\frac{a}{a_\perp} \right)^{-1}. \tag{1.54}$$

The appearance of a resonance term in the denominator is the Confinement Induced Resonance at $a = a_\perp/\mathcal{C}$ [38]. At the CIR, the g_{1D} can be tuned from $-\infty$ to $+\infty$, *i.e.* from strongly attractive to strongly repulsive by varying the transverse width of the waveguide, $a_\perp$, over a small range in the vicinity of the resonance.

Along with the above analytic result, a numerical approach has also been carried out [39] for two different cases: the 6-12 potential,

$$V(r) = \frac{C_{12}}{r^{12}} - \frac{C_6}{r^6}, \tag{1.55}$$

and the spherical square well,

$$V(r) = -V_0\,S(b - r), \tag{1.56}$$

$S(b - r)$ being the unit step function. The results of these numerical calculation are shown in Figure 1.11: they clearly exhibit a singularity in g_{1D}

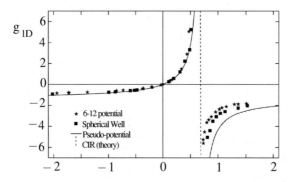

Figure 1.11. The 1D coupling constant, g_{1D}, in units of $\hbar^2/(m_r a_\perp)$, as a function of $a/a_\perp$ for the 6-12 potential (stars) and the spherical well potential (squares), as compared to pseudopotential theory (solid line). From [39].

at $a/a_\perp \simeq 1/\mathcal{C}$. For a less than this value, the calculated g_{1D} values agree well with the analytic expression. For $a/a_\perp > 1/\mathcal{C}$, there are deviations from pseudoptential theory that are greater for the 6-12 potential than for the spherical well. These numerical results validate the pseudopotential analytic result, demonstrating that the CIR is indeed a physical phenomena rather than as artifact of the pseudopotential approximation. It has been shown [39] that CIR is in fact a zero-energy Feshbach resonance, occurring when the energy of a bound state of the asymptotically closed channels (*i.e.* the excited transverse modes) coincides with the continuum threshold of the open channel (lowest transverse mode).

This behavior of 1D system has also been verified experimentally [40].

1.1.4. Luttinger liquid

Fermions in 1D exhibit some exotic properties that cannot be described by the conventional Landau theory of "normal Fermi liquids" [41]. The basic assumption of Landau's theory is that low-energy excitations are quasiparticles with a very long lifetime and can be treated by means of a simple free-energy functional involving deviations from the Fermi ideal-gas momentum distribution. The theory predicts that the low-temperature thermodynamic properties of an interacting Fermi liquid are very similar to those of the non-interacting system: the interactions only lead to quantitative renormalizations. Extension to non-equilibrium properties in a Boltzmann-like viewpoint based on the Vlasov-Landau equation leads to predict the existence of collisionless collective excitations (zero-sound), in addition to ordinary collisional sound, and allows the study of transport phenomena.

The properties of the 1D *Luttinger liquids* of interacting fermions are fundamentally different from those of 2D or 3D normal Fermi liquids [42, 43, 41, 44]. Their elementary excitations are not quasiparticles, but rather collective oscillations of the charge and spin densities, which in general propagate at different speeds giving rise to *spin-charge separation*. Most correlation functions show non-universal power laws with interaction-dependent parameters. Luttinger-liquid behaviour is experimentally well established in the physics of quantum Hall edge states [41] and of quantum spin chains. In contrast, it has been argued that presently available quantum wire systems are not in the regime where Luttinger-liquid effects are important [45]. The formation of a gap in the spin excitation spectrum in the Luther-Emery liquid provides a 1D analog of a superconductor and will be reviewed in Section 1.1.5. Newly obtained results from the application of 1D models to confined gases of Fermi atoms will be presented in Chapters 3 and 4.

In this section our main aim is to trace the line of argument that leads to the exactly soluble Tomonaga-Luttinger model (LM) [46, 47, 48, 49, 50] and then to the concept of Luttinger liquid after inclusion of the spin degree of freedom. We start by considering a homogeneous 1D system of non-interacting spinless fermions, with parabolic dispersion $e_k = k^2/2m$ ($\hbar = 1$). In its ground state the single-particle states with $k < k_F$ are occupied and those with $k > k_F$ are empty, k_F being related to the 1D particle density n by $k_F = \pi n$ (in the spinless case). For low-energy excitations (within a momentum cutoff Λ, say) only the region around the two Fermi points at $\pm k_F$ is involved, and in this region (the "low-energy sector") the dispersion relation can be linearized as $e_{k,\pm} = e_0 \pm v_F k$, with $e_0 = e_F - v_F k_F$. In the LM one lets the cutoff Λ go to infinity. There then are two branches of particles, the right movers R (with positive velocity) and the left movers L (with negative velocity). This modification makes the spinless model exactly soluble even when interactions are switched on.

Let $\hat{\rho}_\alpha(q)$ be the Fourier components of the particle density operator,

$$\hat{\rho}_\alpha(q) = \sum_k \hat{c}^\dagger_{\alpha,k+q} \hat{c}_{\alpha,k} \tag{1.57}$$

for right and left movers, with $\alpha = +$ or $-$ (R or L). The non-interacting Hamiltonian $\mathcal{H}_0$ (and a more general model including interactions, see below) can be written in terms of these operators: in essence, one cannot add a particle (or a hole) to the 1D system without creating at the same time a density wave. A proof of this result is based on the following facts:

(i) the density fluctuation operators obey bosonic commutation relations,

$$[\hat{\rho}_\alpha(-q), \hat{\rho}_{\alpha'}(q')] = \delta_{\alpha\alpha'}\delta_{qq'}\alpha q L/2\pi \qquad (1.58)$$

where L is the system size (special care is needed in calculating the "anomalous commutator" $[\hat{\rho}_\alpha(-q), \hat{\rho}_\alpha(q)]$);

(ii) $\hat{\rho}_\pm(q)$ creates eigenstates of non-interacting Hamiltonian $\mathcal{H}_0$ with energy $\pm q v_F$, so that H_0 can be rewritten as

$$\mathcal{H}_0 = v_F \sum_{q>0,\alpha} \alpha q \, \hat{\rho}_\alpha(q)\hat{\rho}_\alpha(-q) \qquad (1.59)$$

(the linearization of the dispersion relation implies that all electron-hole pairs have the *same energy* $q v_F$ independently of k, so that the states created by the density fluctuation operator are coherent linear combinations of individual electron-hole excitations);

(iii) the spectra of the fermionic and bosonic representations of $\mathcal{H}_0$ are thus the same (one can demonstrate that the degeneracies of the levels are also the same).

The next step in the development of the spinless LM involves the switching on of the interactions starting from the basic interaction Hamiltonian $\mathcal{H}_{int} = \frac{1}{2L} \sum_{q\neq0} v_q \hat{n}_q \hat{n}_{-q}$ where v_q is the Fourier transform of the interparticle potential and $\hat{n}_q$ is the particle density fluctuation operator. There are two basic types of scattering processes contributing to $\mathcal{H}_{int}$: forward scattering (at $q \approx 0$ in the low-energy sector) and backward scattering (at $q \approx 2k_F$ in the low-energy sector), so that in a conventional notation $\mathcal{H}_{int}$ is rewritten as

$$\mathcal{H}_{int} = \frac{1}{2L} \sum_{q\neq0,\alpha} [V_1(q)\hat{\rho}_\alpha(-q)\hat{\rho}_\alpha(q) + V_2(q)\hat{\rho}_{-\alpha}(-q)\hat{\rho}_\alpha(q)]. \qquad (1.60)$$

The couplings are related to the interparticle potential by $V_1(q) = v_q$ and $V_2(q) = v_q - v_{2k_F}$.

The Hamiltonian of the spinless LM in equations (1.59) and (1.60), being expressed solely in terms of density fluctuation operators, can be rewritten as

$$\mathcal{H}_{LM} = \sum_{q\neq0} \left[\left(v_F + \frac{V_1(q)}{2\pi} \right) |q| \hat{b}_q^\dagger \hat{b}_q + \frac{V_2(q)}{4\pi} |q| \left(\hat{b}_q^\dagger \hat{b}_{-q}^\dagger + \hat{b}_{-q}\hat{b}_q \right) \right]$$

$$(1.61)$$

where

$$\hat{b}_q = \sqrt{\frac{2\pi}{L|q|}} \left(\Theta(q)\hat{\rho}_R(q) + \Theta(-q)\hat{\rho}_L(q) \right). \qquad (1.62)$$

The Hamiltonian (1.61) can be diagonalized by a Bogoliubov transformation, with the result

$$\mathcal{H}_{\text{LM}} = \sum_{q \neq 0} \omega_q \hat{\beta}_q^\dagger \hat{\beta}_q \qquad (1.63)$$

where $\omega_q = c_q |q|$ with

$$c_q = \sqrt{\left(v_{\text{F}} + \frac{V_1(q)}{2\pi}\right)^2 - \left(\frac{V_2(q)}{2\pi}\right)^2}. \qquad (1.64)$$

The boson operators $\hat{\beta}_q^\dagger$ and $\hat{\beta}_q$ are linear combinations of the original density fluctuation operators, so that the elementary excitations of the Hamiltonian (1.63) are coherent superpositions of collective density oscillations. Evidently, the stability condition

$$|2\pi v_{\text{F}} + V_1(q)| > |V_2(q)| \qquad (1.65)$$

must be satisfied for the dispersion relation to be real. In addition, the ground-state wave function is normalizable if the stability condition

$$\lim_{q \to \infty} \frac{q^{1/2} V_2(q)}{2\pi v_{\text{F}} + V_1(q)} = 0 \qquad (1.66)$$

is satisfied. Notice that in the non-interacting gas limit one recovers zero sound in the Fermi gas ($\omega_q = v_{\text{F}}|q|$): the electron-hole pairs have been replaced by this collective excitation.

In the case of short-range interactions the dispersion relation (1.64) describes sound waves at long wavelengths, when the coupling parameters tend to constant values ($V_1(q) \to 2g_4$ and $V_2(q) \to g_2$, say). One can prove that in this limit the Hamiltonian (1.63) can be transformed into a continuum-model Hamiltonian describing an elastic string:

$$\mathcal{H}_{\text{LM}} = \frac{1}{2} \int dx \left[\pi u K \, \Pi^2(x) + \frac{u}{\pi K} (\partial_x \phi(x))^2 \right]. \qquad (1.67)$$

Here the fields $\phi(x)$ and $\Pi(x)$, obeying canonical commutation relations $[\phi(x), \Pi(x')] = i\delta(x - x')$, are related to the particle density $\rho(x)$ and to the current density $J(x)$ by $\partial_x \phi(x) = -\pi[\rho(x) - \rho_0]$ and $J(x) = uK\Pi(x)$, ρ_0 being the average particle density in the ground state. The parameters in equation (1.67) are

$$\begin{cases} u = \sqrt{\left(v_{\text{F}} + \dfrac{g_4}{\pi}\right)^2 - \left(\dfrac{g_2}{2\pi}\right)^2} \\[3ex] K = \sqrt{\dfrac{2\pi v_{\text{F}} + 2g_4 - g_2}{2\pi v_{\text{F}} + 2g_4 + g_2}} \end{cases} \qquad (1.68)$$

The form of the transformed Hamiltonian in equation (1.67) emphasizes the collective nature of the particle motions on a chain: a particle setting out with a given momentum will collide with a first neighbor and exchange momentum with it to start a density wave along the chain.

We merely quote without proof at this point three remarkable properties of the quantum fluid described by the LM:

(a) *Momentum distribution.* The momentum occupation number $n_{k,\alpha}$ has the same form for right and left movers and obeys the relation

$$n_{k_F + \delta k, \alpha} = 1 - n_{k_F - \delta k, \alpha}. \qquad (1.69)$$

Thus $n_{k_F, \alpha} = \frac{1}{2}$ (as in the Fermi gas), but the function $n_{k,\alpha}$ is continuous through k_F. However, for any non-vanishing interaction the momentum distribution and the density of states have power-law singularities at the Fermi level, with a vanishing single-particle density of states at ε_F. The absence of a step discontinuity at k_F implies the absence of a quasiparticle pole in the one-particle Green's function.

(b) *Friedel oscillations and charge density waves.* The Friedel oscillations of the displaced electron density around a charged impurity in a normal 1D electron fluid have the form $\cos(2k_F x)/|x|$, where the wavelength $2\pi/(2k_F)$ is equal to the mean first-neighbour distance L/N in the spinless case. The role of repulsive interactions in the LM is to multiply the above factor by a function expressing an extremely slow spatial decay (of course, any form of long-range order such as Wigner crystallization is prevented by fluctuations in 1D according to the Wagner-Mermin theorem). One may rather view the spinless LM fluid as being prone to formation of a charge-density modulation having wave number $2k_F$ and slowly decaying amplitude, with the role of the impurity being that of pinning the phase of such a modulation. Inclusion of the spin-$\frac{1}{2}$ degree of freedom adds to the above result for the screening charge a term having the form $\cos(4k_F x)$ multiplied again by an extremely slow decay factor. Predisposition to charge density waves (CDWs) with wavelength $2\pi/(4k_F) = L/N$ (with k_F being now equal to $\pi n/2$) is indicated in the repulsive Luttinger liquid.

(c) *Electrical conductance.* The current I flowing through a homogeneous 1D fluid subject to a potential drop ΔV applied between a source (at the left) and a drain (at the right) is the difference between the current carried by right movers driven by the source potential towards the drain and the current of left movers driven by the drain potential towards the source. The value of I is determined by the proper current-density – charge-density response function, and in the spinless

LM this function coincides with that of the non-interacting fluid. The result is $I = G\Delta V$ where the conductance G has the fundamental value

$$G = e^2/h \qquad (1.70)$$

(or twice this value if spin is included).

Inclusion of a point-like impurity in the fluid causes a back-scattering current, which is due to the transfer of right movers across the wire to the left-moving branch minus the transfer of left movers to the right-moving branch. The result is a reduction of the net current flowing through the wire at a given value of ΔV. The back-scattering current is proportional to $(\Delta V)^{2g-1}$, where (i) $g = 1$ in the absence of interactions, so that ohmic behaviour is preserved though with a lowered conductance; but (ii) $g < 1$ for repulsive interactions, implying a vanishing conductance for $\Delta V \to 0$. Pinning of the electron density by the impurity is again indicated.

Adding the spin-$\frac{1}{2}$ degree of freedom: spin-charge separation

Up to this point we have ignored the spin of the electron (except for our comment on CDWs and pinning). Upon inclusion of the spin-$\frac{1}{2}$ degree of freedom, all operators in equations (1.59) and (1.60) acquire a spin index σ and summations over σ need adding. However, $2k_F$-scattering of pairs of electrons with antiparallel spins can now also be accompanied by spin flips, and inclusion of this effect makes the model no longer exactly soluble in general (see Section 1.1.5). If this term is neglected, one gets the Luttinger-liquid Hamiltonian in the form

$$\mathcal{H}_{\text{LL}} = v_F \sum_{q>0,\sigma,\alpha} \alpha q \,\hat{\rho}_{\alpha,\sigma}(q)\hat{\rho}_{\alpha,\sigma}(-q)$$

$$+ \frac{1}{2L}\sum_{q\neq 0} V_1(q)\left[\hat{\rho}_R(q) + \hat{\rho}_L(q)\right]\left[\hat{\rho}_R(-q) + \hat{\rho}_L(-q)\right]$$

$$+ \frac{1}{2L}\sum_{q\neq 0,\sigma} [V_2(q) - V_1(q)]\left[\hat{\rho}_{R,\sigma}(-q)\,\hat{\rho}_{L,\sigma}(q) + \hat{\rho}_{L,\sigma}(-q)\hat{\rho}_{R,\sigma}(q)\right],$$

$$(1.71)$$

where $\hat{\rho}_\alpha(q) = \hat{\rho}_{\alpha,\uparrow}(q) + \hat{\rho}_{\alpha,\downarrow}(q)$. One can now carry out a transformation to boson operators, using equation (1.63) with spin-indexed operators, and eliminate the cross terms between charge and spin density by introducing spin-symmetric and spin-antisymmetric combinations,

$$
\begin{cases}
\hat{b}_q^C = \dfrac{1}{\sqrt{2}} \left(\hat{b}_{q\uparrow} + \hat{b}_{q\downarrow} \right) \\[3mm]
\hat{b}_q^S = \dfrac{1}{\sqrt{2}} \left(\hat{b}_{q\uparrow} - \hat{b}_{q\downarrow} \right)
\end{cases}
\tag{1.72}
$$

with the result $\mathcal{H}_{LL} = \mathcal{H}_C + \mathcal{H}_S$ where

$$
\begin{aligned}
\mathcal{H}_C &= \sum_{q \neq 0} \Bigg[\left(v_F + \frac{V_1(q)}{\pi} \right) |q| \hat{b}_q^{C\dagger} \hat{b}_q^C \\
&\quad + \frac{V_2(q) + V_1(q)}{4\pi} |q| \left(\hat{b}_q^{C\dagger} \hat{b}_{-q}^{C\dagger} + \hat{b}_{-q}^C \hat{b}_q^C \right) \Bigg], \\
\mathcal{H}_S &= \sum_{q \neq 0} \left[v_F |q| \hat{b}_q^{S\dagger} \hat{b}_q^S + \frac{V_2(q) - V_1(q)}{4\pi} |q| \left(\hat{b}_q^{S\dagger} \hat{b}_{-q}^{S\dagger} + \hat{b}_{-q}^S \hat{b}_q^S \right) \right].
\end{aligned}
$$

Each of these two independent terms can be diagonalized by a Bogoliubov transformation, finding two branches of excitations: charge density waves propagating with velocity

$$
c_q^C = \sqrt{ \left(v_F + \frac{V_1(q)}{\pi} \right)^2 - \left(\frac{V_2(q) + V_1(q)}{2\pi} \right)^2 }
\tag{1.73}
$$

and spin density waves propagating with velocity

$$
c_q^S = \sqrt{ v_F^2 - \left(\frac{V_2(q) - V_1(q)}{2\pi} \right)^2 }.
\tag{1.74}
$$

Spin-charge separation is implicit in the fact that these two velocities are in general different. Thus, if we write the spin-indexed boson operator as the sum of a charge component and a spin component by inversion of equation (1.72), these two components will evolve in time according to two different and independent Hamiltonians, and will propagate at different velocities.

1.1.5. Electrons in one dimension: spin pairing in the Luther-Emery liquid

The last term in the Hamiltonian (1.71) includes only $2k_F$-scattering between parallel-spin electrons. In their work Luther and Emery [51] added a $2k_F$-scattering term between antiparallel-spin electrons and showed that for special values of the coupling strength parameters exact diagonalization is still possible *via* bosonization. The large-momentum transfer terms are described by these authors through a Hamiltonian H_L in the form

$$\mathcal{H}_L = \sum_{\sigma,\sigma'} \int dx \, \psi_{R,\sigma}^{\dagger}(x)\psi_{L,\sigma'}^{\dagger}(x)\psi_{R,\sigma'}(x)\psi_{L,\sigma}(x)\left[g_{\|}\delta_{\sigma\sigma'} + g_{\perp}\delta_{\sigma,-\sigma'}\right]$$

(1.75)

and one can still carry out a separation of the total Hamiltonian into independent density and spin-density terms. The latter may also be exactly diagonalized with the specific choice $g_{\|} = -3(2\pi v_F)/5$, leading to a new eigenvalue spectrum given by

$$\omega_q = v_F k_F \pm \sqrt{(q \mp k_F)^2 + \Delta^2}$$

(1.76)

where $\Delta = g_{\perp}/(2\pi A)$ with v_F/A the bandwidth. Equation (1.76) shows that an energy gap has opened up at $\pm k_F$ in the spin-density excitation spectrum. There still is no gap in the charge-density excitation spectrum, so that at sufficiently low temperatures only these excitations will be important. The question arises as to whether the gap in the spin excitation spectrum is associated with pairing of antiparallel-spin fermions (see below).

We pursue the presentation of the Luther-Emery liquid by following the discussion of Schulz *et al.* [43], who use the continuum model extended to include the spin-$\frac{1}{2}$ degree of freedom. These authors add a spin-flip back-scattering term in the form

$$\mathcal{H}_{sf} = \frac{g_1}{L} \sum_{k,p,q} \sum_{\sigma,\sigma'} c_{R,\sigma}^{\dagger}(k)c_{L,\sigma'}^{\dagger}(p)c_{R,\sigma'}(p + 2k_F + q)c_{L,\sigma}(k - 2k_F - q).$$

(1.77)

The Hamiltonian is then written in the form

$$\mathcal{H} = \mathcal{H}_C + \mathcal{H}_S + \frac{2g_1}{(2\pi A)^2} \int dx \cos(\sqrt{8}\phi_S(x))$$

(1.78)

where for $i = C, S$ we have

$$\mathcal{H}_i = \frac{1}{2} \int dx \left[\pi u_i K_i \Pi_i^2(x) + \frac{u_i}{\pi K_i}(\partial_x \phi_i(x))^2\right]$$

(1.79)

with

$$\begin{cases} u_i = \sqrt{\left(v_F + \dfrac{g_{4,i}}{\pi}\right)^2 - \left(\dfrac{g_i}{2\pi}\right)^2} \\[2em] K_i = \sqrt{\dfrac{2\pi v_F + 2g_{4,i} - g_i}{2\pi v_F + 2g_{4,i} + g_i}} \end{cases}$$

(1.80)

(see equations (1.67) and (1.68)), having defined $g_C = g_1 - 2g_2$, $g_S = g_1$, $g_{4,C} = g_4$ and $g_{4,S} = 0$. For $g_1 = 0$ equations (1.78)-(1.80) describe independent long-wavelength oscillations of the charge and spin density with linear dispersion relations $\omega_i = u_i|q|$. For $g_1 \neq 0$ the last term on the RHS of equation (1.78) must be treated perturbatively by means of a renormalization-group approach, and the results are as follows.

(i) *Repulsive interactions.* For $g_1 > 0$ the cosine term in equation (1.78) is renormalized to zero at long wavelengths, so that spin-flip processes become irrelevant in the renormalization-group sense. A similar fate is met by corrections to the Hamiltonian (1.78) such as those associated with band curvature and with the absence of high-energy single-particle states. Lattice effects intervene at low energy only to give rise to higher harmonics at wave numbers of the form $q = (2n + 1)k_F$ where n is an integer.

(ii) *Attractive interactions.* In the case $g_1 < 0$, however, the renormalization group scales to strong coupling. In this case the elementary excitation of the spin-fluctuation field ϕ_S may, for instance, correspond to small oscillations around one of the minima of the cosine term, or possibly to soliton-like terms where ϕ_S goes from one minimum to the other. Both types of spin excitation have a finite activation energy: thus for $g_1 < 0$ the spin excitation spectrum has a gap, whereas the charge excitation spectrum remains "massless" (*i.e.* gapless).

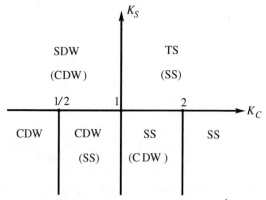

Figure 1.12. Phase diagram of the 1D system of spin-$\frac{1}{2}$ fermions. The spin sector is massless in the upper part ($K_S > 1$), while in the lower part ($K_S < 1$) the spin excitations are massive with a gap Δ.

Figure 1.12 reports from Giamarchi's book the phase diagram for the 1D system of spin-$\frac{1}{2}$ fermions for isotropic spin couplings as assumed so far, in the (K_C, K_S) plane. The plane is divided into four sectors depending on the value of the parameter K_C and of the back-scattering g_1 (or equivalently of the parameter K_S, with the two upper sectors corresponding to $g_1 > 0$ and the two lower sectors to $g_1 < 0$). The indicated phases correspond to the most divergent susceptibility, while the subdominant divergences are indicated in parentheses. The symbols CDW and SDW indicate charge-density-wave and spin-density-wave phases (see Figure 1.13). Also present are singlet superconductivity (SS) and triplet superconductivity (TS).

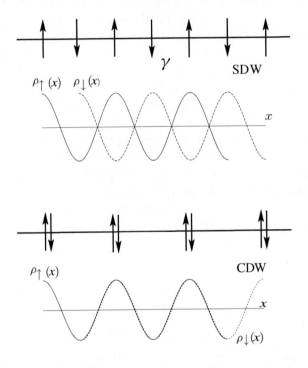

Figure 1.13. Schematic drawing of a spin-density wave (SDW) and a charge-density wave (CDW) with wavelength $2\pi/(2k_F)$. In a SDW one has two waves of spin density for up and down spins, with wavelength $2\pi/(2k_F)$: the two waves are shifted relative to each other by half a wavelength and form antiferromagnetic ordering. In a CDW the two waves are in phase giving a charge-density modulation and no spin-density modulation.

In concluding this section we return to the question of whether spin pairing is present in the Luther-Emery liquid. Seidel and Lee [52, 53] have applied a finite-size bosonization method to study the Luttinger model with large momentum scattering in a state with spin gap and gapless charge degrees of freedom. They calculate the dependence of the ground-state energy on an applied Aharonov-Bohm magnetic flux Φ and show that in the limit of large system size it has an exact period of $hc/(2e)$ corresponding to half a flux quantum, with exponentially small finite-size corrections. In systems containing an even number of particles this periodicity is triggered by the spin gap alone and is independent of whether the superconducting pair-pair correlations are the dominant long-distance/time correlation functions. Their conclusion is that every spin-gapped system with linearly dispersing charge excitations should share some features of a superconductor. Finally, spin pairing in real space inside a 1D system of fermions with attractive interactions will be directly displayed in Chapter 4.

1.2. Experimental realization of quantum gases in quasi-1D geometries

In this section we briefly present the experimental apparatuses such as magneto-optical traps and optical lattices, which are routinely used to create confined quantum gases.

1.2.1. Magneto-optical traps

The principle of laser cooling makes use of the facts that: i) resonant light can exert an enormous force ($\mathcal{O}(10^4)$ times gravity) [54] on atoms and ii) this force depends on the atoms velocity, both its magnitude and direction.

Figure 1.14 illustrates the latter point: an atom moves into a laser beam which is tuned at a frequency slightly below the atomics resonance frequency (red detuning). However the Doppler shift due to the atomics velocity (both in magnitude and direction), makes the atom still experience the laser light as resonant. The atom can then absorb a photon from the laser beam and emit it in all directions. Each absorbed photon delivers a small momentum kick to the atom in the laser-beam direction *i.e.* opposite to the atom's direction of motion, which the emitted photons deliver momentum kicks in all directions: the net result is that the atom's motion in the direction of the laser beam is reduced. Figure 1.14 illustrates that when the atom's velocity in the laser beam direction becomes smaller, the Doppler shift becomes smaller and the induced laser force gets smaller. Figure 1.14 (c) shows an atom moving in between two laser beams and the force indicated here is now slightly smaller than the force shown in Figure 1.14 (a): the laser beam in the direction of the atom's motion also exerts a small force in the opposite direction.

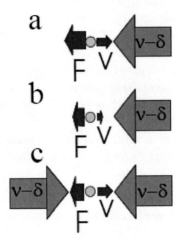

Figure 1.14. The light forces on an atom depending on its velocity (laser light detuned by δ from the resonance frequency v). From [55].

By extension to three pairs of laser beams one can cool the motions of all atoms in all directions.

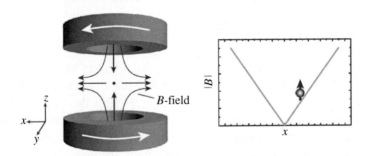

Figure 1.15. Quadrupole coil pair for generating a magnetic quadrupole trap. Two coils carry opposite currents and form a magnetic field which vanishes in the trap center and increase linearly in each direction. From [56].

To create a trap, the light forces must also be dependent on the atom's position with respect to the trap center. This can be achieved with the installation of an anti-Helmholtz magnetic field (like the Helmholtz configuration but with the current through one coil reversed). The magnetic field of such a configuration is zero in the center but increases in all di-

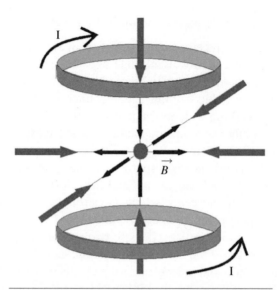

Figure 1.16. Geometry of a magneto optical trap (MOT). The quadrupole field of the MOT is generated by an anti Helmholtz coil pair. Laser beams with circular polarized light from six directions are overlapped to the trap center. From [57].

rections [56]. From this point, the absolute value of the magnetic field increases linearly in each direction as $B(\mathbf{r}) = \sqrt{4x^2 + y^2 + z^2}$. Atoms outside the center will find themselves in a magnetic field and the Zeeman effect will then change the resonance frequency of the atomic transition because the ground and excited states are affected differently by the magnetic field. The inhomogeneous magnetic field together with the six red-detuned laser beams of proper polarization (Figure 1.16) make the magneto-optical trap (MOT). When an atom is placed in an inhomogeneous magnetic field $B(\mathbf{r})$ it is subjected to an external force directed toward the magnetic field minimum or maximum depending on the orientation of the magnetic moment.

When an atom moves in such a field, it maintains the relative orientation of the magnetic moment if the change of the direction of the magnetic field is small compared to the Larmor frequency. An atom with magnetic quantum number m_F and Landé g-factor g_F in the hyperfine state F is therefore subjected to a potential formed by the Zeeman energy shift

$$V_{mag}(\mathbf{r}) = -g_F \, m_F \, \mu_B \, B(\mathbf{r}), \tag{1.81}$$

where μ_B is the Bohr magneton. In the center of MOT, the magnetic field cancels between two coils that carry opposite currents.

1.2.2. Linear standing waves and optical lattices

Optical lattices have received much attention in atomic physics because atoms cooled to the microkelvin range and interacting with an optical standing wave (schematic setup in Figure 1.17) are bound in the antinodes of the standing wave, thus forming a regular microscopic lattice [58]. One of the goals of working with optical lattices is to achieve atomic densities such that the number of atoms trapped in the lattice exceeds the number of lattice sites. Several difficulties have to be overcome to achieve such high densities: (i) one needs a source of atoms that provides densities large enough to fill each lattice site, and (ii) it is also necessary to have a lattice in which fluorescence effects are strongly suppressed such that the optical trapping potential is hardly influenced by the large number of particles present in the lattice. If these two problems can be overcome it is possible to study the effects of quantum statistics of the particles trapped in the lattice and to investigate the effects of their interaction. As has been shown experimentally [58], problem (ii) can be overcome by using optical lattices that are far-blue detuned with respect to an atomic transition.

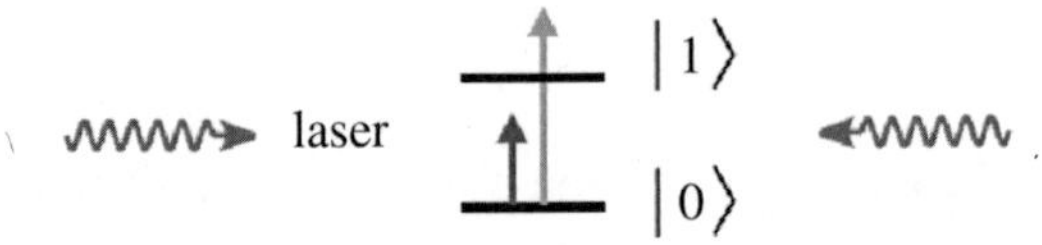

Figure 1.17. Schematic setup of an optical lattice configuration. From [59].

The basic ideas are as follows. When an atom is placed into a laser-light field, the electric field $\mathbf{E}$ (oscillating with the complex amplitude E at a frequency $\omega = 2\pi\nu$) induces an oscillating dipole moment μ_d. For a single atom, the Hamiltonian can be written as

$$\mathcal{H}_A = \frac{P^2}{2m} + \sum_i \hbar\,\omega_i |e_i\rangle\langle e_i|. \tag{1.82}$$

Here, P is the center-of-mass momentum operator and $|e_i\rangle\langle e_i|$ specifies the internal atomic states with energy ω_i. We assume the atom to initially occupy an internal state $|e_0\rangle \equiv |a\rangle$, which defines the point of zero energy. The atom is subject to a laser field with electric field

$$\mathbf{E}(x, t) = E(x, t) \exp(-i\omega t)\hat{\epsilon}, \tag{1.83}$$

where ω and $\hat{\epsilon}$ indicates respectively the frequency and the polarization vector. The electric field amplitude $E(x, t)$ is slow-varying in time t

compared to $1/\omega$ and also slow in space x compared to the size of the atom. Under these conditions, the interaction between atom and laser is described in the dipolar approximation by the Hamiltonian

$$\mathcal{H}_{\text{dip}} = -\mu_d E(x,t) + \text{h.c},\qquad(1.84)$$

where μ_d is dipolar operator of the atom. The laser is assumed to be far detuned from any optical transition so that no significant population is transferred from $|a\rangle$ to any of the other internal atomic states *via* $\mathcal{H}_{\text{dip}}$. The additional atomic levels can be treated then in perturbation theory and eliminated from the dynamics. In doing so [59], the AC-Stark shift of the internal state $|a\rangle$ is found in the form of a conservative potential $V(x)$ whose strength is determined by the atomic dipole operator and by the laser light at the center-of-mass position x of the atom. In particular, $V(x)$ is proportional to the laser intensity $|E(x,t)|^2$. Under these conditions, the motion of the atom is governed by the Hamiltonian

$$\mathcal{H} = \frac{P^2}{2m} + V(x).\qquad(1.85)$$

Now we specialize to the case where the dominant contribution to the optical potential arises from one excited atomic level $|e\rangle$ only. In a frame rotating with the laser frequency the Hamiltonian of the atom is approximately given by

$$\mathcal{H} = \frac{P^2}{2m} + \delta|e\rangle\langle e|,\qquad(1.86)$$

where $\delta = \omega_e - \omega$ is the detuning of the laser from the atomic transition $|e\rangle \leftrightarrow |a\rangle$. The dominant contribution to the atom-laser interaction neglecting all quickly oscillating terms (*i.e.* in the rotating-wave approximation) is given by

$$\mathcal{H}_{\text{dip}} = \frac{\Omega(x)}{2}|e\rangle\langle a| + \text{h.c.}\qquad(1.87)$$

Here

$$\Omega = -2E(x,t)\,\langle e|\mu_e|a\rangle,\qquad(1.88)$$

is the so-called Rabi frequency driving the transitions between the two atomic levels. For large detuning $\delta \gg \Omega$, adiabatically eliminating the level $|e\rangle$ yields the following explicit expression for the optical potential:

$$V(x) = \frac{|\Omega(x)|^2}{4\delta}.\qquad(1.89)$$

Furthermore, if the atom is interacting with a standing light wave, the spatial dependence of the Rabi frequency is given by [59]

$$\Omega(x) = \Omega_0 \sin(kx). \tag{1.90}$$

Thus, according to the equation (1.89), the periodic potential can be written as

$$V(x) = V_0 \sin^2(kx), \tag{1.91}$$

where $V_0 = \Omega_0^2/4\delta$ denotes the potential depth of the lattice and is usually expressed in natural units of the recoil energy $E_r = \hbar^2 k^2/(2m)$. This optical potential is periodic in space and it is thus useful to work out the Bloch eigenstates

$$\phi_q^{(n)}(x) = e^{iqx} u_q^{(n)}(x), \tag{1.92}$$

where q is the Bloch wave number and $u_q^{(n)}(x)$ are the eigenstates of the Hamiltonian

$$\mathcal{H}_q = \frac{(p+q)^2}{2m} + V_0(x) \tag{1.93}$$

with energy $E_q^{(n)}$ and periodicity a of the optical potential, *i.e.*

$$\mathcal{H}_q u_q^{(n)}(x) = E_q^{(n)} u_q^{(n)}(x). \tag{1.94}$$

Usually the particles are taken to be in the lowest band, which implies cooling to temperatures T much lower than the trapping frequency ω_T.

A set of orthogonal normalized wave functions that fully describe particles in band n of the optical potential and are localized at the lattice sites (in regions around the potential minima) is given by the Wannier functions

$$\omega_0(x - x_i) = \mathcal{N}^{-\frac{1}{2}} \sum_q e^{-iqx_i} \phi_q^{(n)}(x) \tag{1.95}$$

where x_i is the position of the lattice site and $\mathcal{N}$ is a normalization constant. The advantages of using Wannier functions $\omega_0(x - x_i)$ to describe particles in the lowest band are that

(i) a mean position x_i may be attributed to the particle if it is found to be in the mode corresponding to the Wannier function $\omega_0(x - x_i)$;

(ii) local interactions between particles can be described easily since the dominant contribution comes from particles located at the same position x_i.

1.3. Quantum phase transitions in ultracold atomic gases

The main aim of this section is to present a formal mapping between the many-body Hamiltonian of a confined quantum gas in a strong optical lattice and the so-called inhomogeneous Hubbard model. The bosonic version of the Hubbard model and its phase diagram, with particular attention to the Mott superfluid-insulator quantum phase transition and its observation [60], are presented in this section while the fermionic version is extensively discussed in Chapter 3.

1.3.1. The Hubbard model for ultracold atoms in optical lattices

Bosons at $T = 0$ in a periodic potential (*e.g* inside an optical lattice) exhibit two types of phases: a superfluid phase and a Mott insulating phase characterized by integer (commensurate) boson density [61]. The simplest system where one can investigate the Mott-insulator phase transition occurring in these systems at zero temperature is a gas of repulsively interacting bosons with spin zero in a lattice described by the Bose-Hubbard (BH) Hamiltonian. This Hamiltonian contains the main physics of strongly interacting Bose systems which is the competition between kinetic and interaction energy.

The starting point is the Hamiltonian $\mathcal{H}_{\text{full}}$ of many interacting bosons [62],

$$
\begin{aligned}
\mathcal{H}_{\text{full}} = & \int d^3x \, \psi^\dagger(\mathbf{x}) \left[-\frac{\hbar^2}{2m} \nabla^2 + V_0(\mathbf{x}) + V_T(\mathbf{x}) \right] \psi(\mathbf{x}) \\
& + \frac{g}{2} \int d^3x \, \psi^\dagger(\mathbf{x}) \psi^\dagger(\mathbf{x}) \psi(\mathbf{x}) \psi(\mathbf{x}),
\end{aligned}
\tag{1.96}
$$

where $\psi(\mathbf{x})$ is a boson field operator for atoms in a given internal atomic state $|0\rangle$, $V_0(\mathbf{x})$ is the optical lattice potential, and $V_T(\mathbf{x})$ describes an external trapping potential, *e.g* a magnetic trap (slowly varying compared to $V_0(x)$). As mentioned before (see equation (1.91)), in the simplest case the 3D optical lattice potential has the form

$$
V_0(\mathbf{x}) = \sum_{j=1}^{3} V_{j0} \sin^2(kx_j).
\tag{1.97}
$$

All the particles are assumed to be in the lowest band of the optical lattice.

The field operator can be expanded in terms of Wannier functions as

$$
\psi(\mathbf{x}) = \sum_i a_i \, \omega_0(\mathbf{x} - \mathbf{x}_i),
\tag{1.98}
$$

where $\hat{a}_i$ is the destruction operator for a particle on site $\mathbf{x}_i$ and $\omega_0(\mathbf{x} - \mathbf{x}_i)$ is the three-dimensional version of the Wannier functions. $\mathcal{H}_{\text{full}}$ can then be written as [62]

$$\mathcal{H}_{\text{full}} = -\sum_{i,j} J_{ij}\hat{a}_i^\dagger \hat{a}_j + \frac{1}{2}\sum_{i,j,k,l} U_{ijkl}\hat{a}_i^\dagger \hat{a}_j^\dagger \hat{a}_k \hat{a}_l, \qquad (1.99)$$

where

$$J_{ij} = -\int dx\,\omega_0(\mathbf{x} - \mathbf{x}_i)\left(\frac{p^2}{2m} + V_0(\mathbf{x}) + V_T(\mathbf{x})\right)\omega_0(\mathbf{x} - \mathbf{x}_j), \quad (1.100)$$

and

$$U_{ijkl} = g\int dx\,\omega_0(\mathbf{x} - \mathbf{x}_i)\omega_0(\mathbf{x} - \mathbf{x}_j)\omega_0(\mathbf{x} - \mathbf{x}_k)\omega_0(\mathbf{x} - \mathbf{x}_l). \quad (1.101)$$

Now, if we include only the interaction matrix element U_{0000} for particles in the same site and the hopping matrix element J_{01} between nearest neighbors we end up by standard Bose-Hubbard Hamiltonian [62]:

$$\mathcal{H}_{\text{BH}} = -J\sum_{<i,j>} \hat{a}_i^\dagger \hat{a}_j + \frac{1}{2}U\sum_i \hat{n}_i(\hat{n}_i - 1) + \sum_i \epsilon_i\,\hat{n}_i \qquad (1.102)$$

where $J \equiv J_{01}$ and $U \equiv U_{0000}$. The operator $\hat{n}_i = \hat{a}_i^\dagger \hat{a}_i$ counts the number of bosonic atoms at lattice site i; the annihilation and creation operators $\hat{a}_i$ and $\hat{a}_i^\dagger$ obey the canonical commutation relation $[\hat{a}_i, \hat{a}_i^\dagger] = \delta_{ij}$. The energy offset ϵ_i of each lattice site arises from the trapping potential and is given by

$$\epsilon_i = \int d^3x\,V_T(\mathbf{x})|\omega_0(\mathbf{x} - \mathbf{x}_i)|^2 \approx V_T(x_i) \qquad (1.103)$$

In summary, the Bose-Hubbard Hamiltonian, equation (1.102), consists of three terms [56]:

i) the hopping term describes the tunneling of bosons between neighboring potential wells and tends to *delocalize* each atom over the lattice;

ii) the interaction of n atoms, each interacting with $n - 1$ other atoms on the same lattice site, is described by the second term of the Bose-Hubbard Hamiltonian and tends to *localize* atoms on lattice sites; and

iii) an external confinement term which gives rise to an energy offset on the i-th site (for a homogeneous system ϵ_i is zero).

In the following section, we mention the possible ground states for this model and we will see how changing the parameters in Hamiltonian can lead to different regimes for this model.

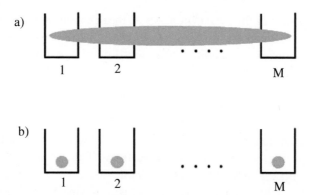

Figure 1.18. Visualization of a) the superfluid phase and b) the Mott-insulator phase with commensurate filling predicted by the Bose-Hubbard model. From [59].

1.3.2. The Bose-Hubbard model and the superfluid-insulator transition

The Bose-Hubbard Hamiltonian, equation (1.102), with repulsive interaction ($U > 0$) has two distinct ground states depending on the strength of U relative to the tunnel-coupling J.

For $U \ll J$, the tunneling term dominates the Hamiltonian and atom-atom interactions do not play an important role. The system is in the weakly interacting regime, where the bosons tend to form a BEC and also tend to delocalize to save kinetic energy [63]. The ground state of the many-body system with N atoms is given by a product of identical single-particle Bloch waves, where each atom is spread over the entire lattice with M lattice sites [64]:

$$|\Psi_{\text{SF}}\rangle_{\frac{U}{J}\approx 0} \propto \left(\sum_{i=1}^{M} \hat{a}_i^\dagger \right)^N |0\rangle \approx \prod_{i=1}^{M} |\phi\rangle_i \qquad (1.104)$$

The atom number in each lattice site then fluctuates forming a poissonian distribution [60]. The non-vanishing expectation value of $\psi_i = \langle \phi_i | \hat{a}_i | \phi_i \rangle$ characterizes the coherent matter-wave field on the i-th lattice site: each wave field has a fixed phase relative to all other coherent wave fields on different lattice sites.

On the other hand, by increasing the repulsive interaction U compared to J a quantum phase transition (at temperature $T = 0$) from the superfluid to a strong correlated phase takes place. In the regime $U \gg J$ the interactions dominate the behavior of the Hamiltonian so that fluctuations in the atom number on each lattice site become energetically costly and the ground state of the system consists of localized atomic wavefunctions that minimize the interaction energy. In this limit, the ground state of the many-body system for a commensurate filling of n atoms per lattice site is given by

$$|\Psi_{\mathrm{MI}}\rangle_{J\approx 0} \propto \prod_{i=1}^{M} (\hat{a}_i^\dagger)^n |0\rangle. \qquad (1.105)$$

In such a situation the atom number on each lattice site is ideally exactly determined but the phase of the matter-wave field on a lattice site has maximum uncertainty. The state is characterized by a vanishing of the matter-wave field on the i-th lattice site, $\psi_i = \langle \phi_i | \hat{a}_i | \phi_i \rangle \approx 0$.

The phase diagram of the Bose system with superfluid and Mott insulator phases is conveniently given in the grand canonical ensemble. The superfluid state is characterized by a finite-range phase coherence with a finite expectation value for the field operator $\langle \hat{a}_i \rangle \neq 0$, while in the Mott insulator $\langle \hat{a}_i \rangle = 0$ and the gas is incompressible. A qualitative phase diagram was already derived in [61]. Figure 1.19 shows a portion of the phase diagram for a macroscopic system on a square lattice, calculated with an analytic strong-coupling series [65]. The diagram shows the boundary between the Mott-insulating and the superfluid phase as a function of the chemical potential μ and the tunnel coupling J, both in units of the on-site interaction U. The two lobes represent a parameter

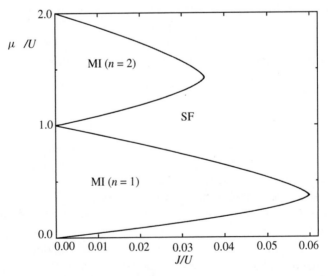

Figure 1.19. Superfluid-Mott insulator phase diagram on a square lattice with 4 next neighbors. From [65].

range for which the ground state is a Mott insulator with an atom number of one and two particles per lattice site, respectively.

For larger chemical potential further lobes representing Mott insulator states with higher atom numbers can be found. For larger n the critical point scales approximately as $(J/U)_c \sim 1/n$. Figure 1.20 shows atomic distribution and interfrence pattern of the two phases.

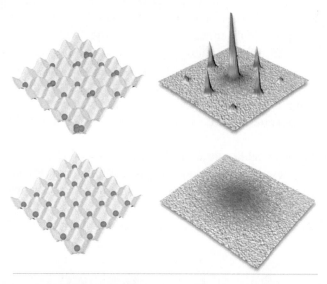

Figure 1.20. (Top Panel) In the superfluid state a Bose-Einstein condensate can be described as a macroscopic matter wave. When the condensate is released from the periodic potential, a matter-wave interference pattern is formed owing to the phase coherence between the atomic wavefunctions on different lattice sites. In this case, the phase of the macroscopic matter wave is well defined, but the atom number on each lattice site fluctuates. (Bottom Panel) In a Mott insulating state of matter, each lattice site is filled with a given number of atoms but the phase of the matter-wave field remains uncertain. No interference can be seen in this case when the quantum gases are released from the lattice potential. From [64].

1.3.3. The observation of the superfluid-insulator transition

Following a sugestion by Jaksch *et al.* [62] a transition between the two states discussed above was observed [60]. In the experiment, a Bose-Einstein condensate of ^{87}Rb atoms is loaded into a 3D optical lattice potential. The system is characterized by a low atom occupancy per lattice site, of the order of $\bar{n}_i \approx 1 - 3$, and thus provides a testing ground state for the Bose-Hubbard model. The crucial parameter U/J that characterizes the strength of the interactions relative to the tunnel coupling between neighboring sites can be varied by simply changing the depth of the optical lattice potential: on increasing the lattice depth, U increases almost linearly due to the tighter localization of the atomic wave packets on each lattice site and J decreases exponentially due to the decreasing tunnel. The ratio U/J can therefore be manipulated [63] over a wide range, from $U/J \approx 0$ up to values of $U/J \approx 2000$. In the superfluid regime, phase coherence of the matter-wave field across the lattice can be revealed by suddenly turning off all trapping fields, so that the individual fields on different lattice sites expand and interfere with each other. After a fixed time-of-flight period, the atomic density distribution can then be measured by absorption imaging: the image directly reveals the momentum distribution of the trapped atoms. In *e.g.* Figure 1.21(b) an interference pattern can be seen after releasing the atoms from a 3D lattice potential.

If, on the other hand, the optical lattice potential depth is increased so that the system is in the Mott insulating regime, phase coherence is lost between the matter-wave fields on neighboring lattice sites. In this case, no interference pattern can be seen in the time-of-flight images. (Figure 1.21 (h))

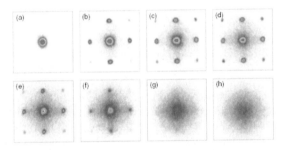

Figure 1.21. Absorption images of matter-wave interference patterns after releasing the atoms from an optical lattice potential with different depths of (a) $0E_r$; (b) $3E_r$; (c) $7E_r$; (d) $10E_r$; (e) $13E_r$; (f) $14E_r$; (g) $16E_r$ and (h) $20E_r$. From [64].

1.3.4. BCS-BEC crossover in fermion superfluid

As described in previous sections, the behaviors of a dilute gases of bosons and fermions are quite different when the phase space density becomes of order unity, *i.e.* when the matter wavelength of the particles is of comparable size to their spacing. While bosons undergo a phase transition into a superfluid state, fermions stack up in a Fermi sea. Creating a superfluid from fermionic matter most often requires pairing of

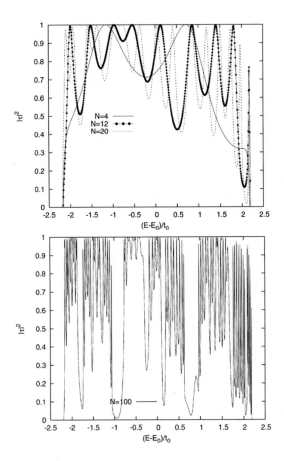

Figure 2.9. Transmission coefficient as in Figure 2.8, for a Fibonacci-ordered array.

discuss about transport of matter by a Bose-Einstein condensate and by a spin-polarized Fermi gas.

2.4.1. Linear standing-wave arrays

A single-well linear lattice for an atomic gas can be realized in the laboratory by superposing on a highly elongated magnetic trap two optical laser beams which counterpropagate along the z axis, say. The standing optical wave acts on the atoms as a periodic array of potential wells, as is shown on the left in Figure 2.10. The potential has the form

$$U(z) = U_0 \sin^2(k_L z), \tag{2.19}$$

where U_0 is the well depth, which is controlled through the laser intensity, and k_L is the laser wavenumber determining the distance d between adjacent wells as $d = \pi/k_L$. The 1D optical lattice can be modified by means of auxiliary laser beams, and in particular its period can be doubled by adding two beams that are rotated with respect to the z axis by angles of 60° and 120°(see bottom part of Figure 2.10). For a suitable choice of phases the potential seen by the atoms takes the shape

$$U(z) = U_0[\sin^2(k_L z) + \delta^2 \sin^2(k_L z/2)], \qquad (2.20)$$

where δ^2 is the relative potential energy difference between adjacent wells. As we have illustrated in Section 2, in solid-state terminology the doubling of the period causes the opening of a minigap in the energy spectrum as a function of δ^2 (see Figure 2.4).

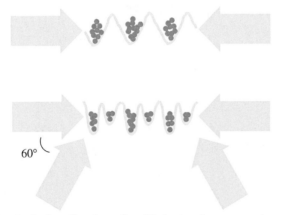

Figure 2.10. Optical realization of a 1D lattice for an atomic gas (top) and doubling of its period (bottom).

A schematic drawing of a set-up of optical lasers that would create an atomic Fibonacci wave guide is shown in Figure 2.11. Here, two pairs of counterpropagating laser beams create a square optical lattice, and the projection of this lattice on a line at an angle $\alpha = \arctan(2/(\sqrt{5} + 1))$ creates a quasi-periodic sequence of bond lengths obeying the Fibonacci rule [109]. The atoms can be made to travel along the sequence by pointing along this direction a hollow beam (for a description of the latter see for example the work of Xu et al. [115]). With this set-up the hopping energies follow the Fibonacci sequence, but in order to make a direct comparison with transport in a double-well lattice we consider in the following the case in which the site energies with relative energy difference δ^2 form the Fibonacci chain.

Finally, we remark that a simple way to create and control a constant external drive on the atomic gas in a 1D lattice would be to tilt the whole set-up by an angle β relative to the vertical axis (see Figure 2.11 for the

Fibonacci array). In this case the external force acting on atoms of mass m in gravity g is $F = mg \cos \beta$.

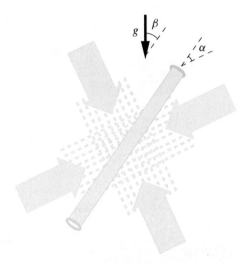

Figure 2.11. Schematic representation of a five laser-beam configuration to create a quasi-1D Fibonacci array for an atomic gas. Two pairs of beams generate a square optical lattice and a hollow beam confines the atoms to a strip with slope $\alpha = \arctan(2/(\sqrt{5} + 1))$ relative to an axis of the square lattice. The angle β between the hollow beam and the vertical direction determines the driving force in the gravity field as $F = mg \cos \beta$.

2.4.2. The local density-of-state

The case of practical interest for atomic gas in 1D optical lattice is that in which a few hundred wells are occupied. The appropriate DOS in the absence of external force is therefore as shown in Figure 4. A constant external force causes a linear tilt of the spectrum and the atoms travel through the lattice and at the same time are accelerated through the energy spectrum [116]. The schematic trajectory of an atom at given initial energy moving in space and through the spectrum is indicated by arrows in Figure 2.12. In this figure we have plotted the local (site-projected) DOS $D_i(E)$ as a function of position and energy for a single-well, a double-well, and a Fibonacci array composed of 100 wells subjected to a force $F \simeq 0.01\, mg$, the local DOS being defined by the expression

$$D_i(E) = -\frac{1}{\pi} \operatorname{Im} \langle i|(E - \mathcal{H})^{-1}|i\rangle = \frac{1}{\pi} \operatorname{Im} G_{i,i}(E). \qquad (2.21)$$

Size effects should be expected in the local DOS only in the region of edge sites.

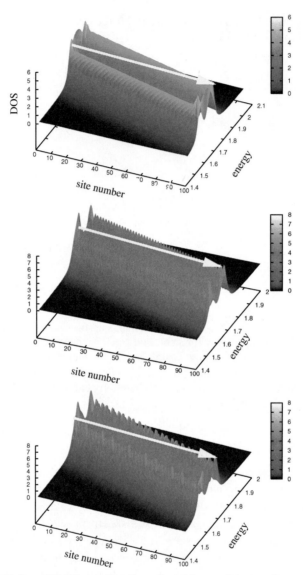

Figure 2.12. Local DOS as a function of position and energy for a single-well, a double-well, and a Fibonacci array.

The Kirkman-Pendry relation in equation (2.3) applies only to the evaluation of the global DOS and other methods must be used for the local DOS. The matrix element $G_{i,i}(E) = [E - \tilde{\varepsilon}_i(E)]^{-1}$ can be evaluated by reducing the array to an effective site with energy $\tilde{\varepsilon}_i(E)$ through decimation of all $j - th$ sites with $j \neq i$. The renormalized Hamiltonian is a c-number and can be expressed in terms of continued fractions [102].

The site-projected DOS in Figure 2.12 shows that at a given energy the population of a sub-set of sites is favored in both the double-well and the Fibonacci array, these sites being rather irregular distributed along the array in the later case [110].

2.4.3. The transmittivity

The transmittivity of an atomic gas driven by a constant force through an optical array can be calculated by an evaluation of the scattering matrix which explicitly takes into account the presence of the bias [117, 108]. For this purpose, the system is connected to incoming and outgoing leads, which mimic its coupling to the continuum by injecting and extracting a steady-state particle current, respectively (see Figure 13). In an actual experiment on a Bose-Einstein condensate such as that carried out by Anderson and Kasevich [96], the incoming lead would correspond to a continuous replenishing of the condensate in the array and the outgoing lead to a particle detection system counting the atoms that leave the array.

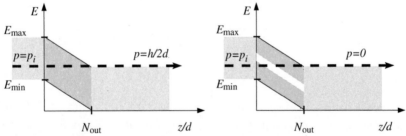

Figure 2.13. Band tilting in space for a single-well array (left) and a double-well array (right), due to a constant force F. Coupling to an incoming and an outgoing lead is also shown for atoms with initial momentum p_i and exiting after N_{out} hops at the Brillouin zone boundary (left) or at the Brillouin zone center (right). The position z is in units of d.

We define $N_{out} = (E_{max} - E_i)/(Fd)$ (with $N_{out} \leq N$) as the number of hops after which an atom having initial energy E_i and momentum p_i reaches the highest energy point in the dispersion curve and can either leave the lattice or be Bragg reflected (see Figure 2.13). In a Bose-Einstein condensation all bosons are initially at the lowest energy E_{min}, namely at $p_i = 0$. In the case of a Fermi gas all fermions with $p_i \in [0, p_F]$, $p_F = (2mE_F)^{1/2}$ being the Fermi momentum, move coherently in the absence of dissipative scattering processes. The leads are again taken to have a band width $4t_0$ and their centers are relatively shifted by $FN_{out}d$ in order to optimize their coupling with the lattice.

The outgoing wavefunction $|\varphi_{out}^{\tilde{\kappa}}\rangle$ is an eigenstate of the outgoing lead and is related to the incoming wavefunction $|\varphi_{in}^{\kappa}\rangle$, which is an eigenstate of the incoming lead, by

$$|\varphi_{\text{out}}^{\tilde{\kappa}}\rangle = (1 + G^0 T)|\varphi_{\text{in}}^{\kappa}\rangle . \tag{2.22}$$

The wavevectors κ and $\tilde{\kappa}$ are uniquely identified by the relations

$$\kappa, \tilde{\kappa} = \frac{1}{d} \arccos\left(\frac{E - E_0^{\text{in,out}}}{2t_0}\right) . \tag{2.23}$$

In equation (2.22) $G^0 = (E - \mathcal{H}_0)^{-1}$ is the Green's function of the disconnected leads described by the Hamiltonian $\mathcal{H}_0 = \mathcal{H}_{\text{L,l}} + \mathcal{H}_{\text{L,r}}$, and the scattering matrix $T(E) = \mathcal{H}_{\text{I}}(1 - G^0\mathcal{H}_{\text{I}})^{-1}$ is referred to the perturbation Hamiltonian

$$\mathcal{H}_{\text{I}} = \tilde{\mathcal{H}}_{\text{w}}|_{N=N_{\text{out}}} - (E_0^{\text{in}}|1\rangle\langle 1| + E_0^{\text{out}}|N_{\text{out}}\rangle\langle N_{\text{out}}|), \tag{2.24}$$

$\tilde{\mathcal{H}}_{\text{w}}|_{N=N_{\text{out}}}$ being given by equation (2.4) for the case $N = N_{\text{out}}$.

Since the wavefunctions $|\varphi_{\text{in}}^{\kappa}\rangle$ and $|\varphi_{\text{out}}^{\tilde{\kappa}}\rangle$ are defined in disconnected spaces, the projection of $|\varphi_{\text{out}}^{\tilde{\kappa}}\rangle$ onto the localized function $|n\rangle$ can be written as

$$\begin{aligned}
\langle n \,|\, \varphi_{\text{out}}^{\tilde{\kappa}}\rangle &= \sum_{j\leq 1,l} G_{n,l}^0 T_{l,j}\langle j|\varphi_{\text{in}}^{\kappa}\rangle \\
&= \sqrt{2}G_{n,N_{\text{out}}}^0 T_{N_{\text{out}},1}\sin(\kappa d)u_{\kappa}(d) ,
\end{aligned} \tag{2.25}$$

where we have set $\langle 1\,|\varphi_{\text{in}}^{\kappa}\rangle = u_{\kappa}(d)(e^{i\kappa d} - e^{-i\kappa d})/(i\sqrt{2})$ with $u_{\kappa}(d)$ being the Wannier function in the potential $U(z)$. The Green's function element $G_{n,N_{\text{out}}}^0$ in equation (2.25) determines the coherence between site n and site N_{out} on the chain for the outgoing lead and can be written as

$$G_{n,N_{\text{out}}}^0 = \frac{t_0^{n-N_{\text{out}}}}{|t_0^{n+1-N_{\text{out}}}|}e^{i\tilde{\kappa}(n+1-N_{\text{out}})d} . \tag{2.26}$$

In an out-of-equilibrium picture, the velocities v_{in} and v_{out} of the incoming and outgoing wavefunctions enter the definition of the transmission coefficient $\mathcal{T}$ as

$$\mathcal{T} = |\tau|^2\frac{v_{\text{out}}}{v_{\text{in}}} = \frac{\lim_{n\to+\infty}\langle n|\varphi_{\text{out}}^{\tilde{\kappa}}\rangle\langle\varphi_{\text{out}}^{\tilde{\kappa}}|n\rangle v_{\text{out}}}{\lim_{m\to-\infty}\langle m|\varphi_{\text{in}}^{\kappa}\rangle\langle\varphi_{\text{in}}^{\kappa}|m\rangle v_{\text{in}}} \tag{2.27}$$

and finally we obtain

$$\mathcal{T} = 4\frac{|T_{1,N_{\text{out}}}|^2}{t_0^2}\sin(\kappa d)\sin(\tilde{\kappa}d). \tag{2.28}$$

Let us remark that equation (2.28) reduces to the transmission coefficient evaluated by means of equation (2.18) if $E_0^{\text{out}} = E_0^{\text{in}}$.

2.4.4. Numerical results for a Bose-Einstein condensate

The particles in a Bose-Einstein condensate can freely tunnel through the optical array if the barrier U_0 is set too high, and in this case the gas behaves as a superfluid with long-range phase coherence. According to the indeterminacy principle, suppression of phase fluctuations implies large fluctuation in the number of bosons at each site of the array. This also implies that the total number $\mathcal{N}_b$ of bosons is much larger than the number N of sites. The Hamiltonian for the condensate can be written as in equation (2.1), taking into account mean-field interaction effects in the evaluation of the site energies and hopping energies.

With the aim to determine the parameters e_i and $t_{i,i+1}$, we proceed to a 1D reduction of the Hamiltonian by introducing the transverse width $\sigma_\perp$ of the boson density in an elongated cigar-shaped harmonic trap and the boson wavefunction in the $i - th$ cell. In a tight-binding scheme $\Psi_i(z)$ is a Wannier function in the potential $U(z)$ and, according to the early work of Slater [118], can be written as a Gaussian function with a width σ_{z_i} determined by the harmonic approximation to the lattice well. That is,

$$\Psi_i(z) = \frac{\Psi_i(0)}{\pi^{1/4}\sigma_{z_i}} \exp[-(z - z_i)^2/(2\sigma_{z_i}^2)] \qquad (2.29)$$

where $|\Psi_i(0)|^2 \sim \mathcal{N}_b/N$ is the number of bosons in the i-th well. The determination of the widths is carried out variationally [119, 108]. The site energies are then given by

$$e_i = \int dz\, \Psi_i(z) \left[-\frac{\hbar^2\nabla^2}{2m} + U(z) + maz + \frac{1}{2}g_{bb}|\Psi_i(z)|^2 + C \right] \Psi_i(z).$$
$$(2.30)$$

Here, $a = F/m$ is the acceleration due to a constant external force F acting on the bosons, $C = \hbar^2/(2m\sigma_\perp^2) + \frac{1}{2}m\omega_\perp^2\sigma_\perp^2$ is a factor taking into account the reduction to 1D dimensionality, and $\omega_\perp$ is the radial frequency of the harmonic trap. Finally, the parameter g_{bb} is the strength of the 1D boson-boson repulsion and is given by [119]

$$g_{bb} = \frac{4\pi\hbar^2}{m}\frac{a_{bb}}{2\pi\sigma_\perp^2} \qquad (2.31)$$

with a_{bb} the boson-boson scattering length. The hopping energies $t_{i,i+1}$ are similarly given by

$$t_{i,i+1} = \int dz\, \Psi_i(z) \left[-\frac{\hbar^2\nabla^2}{2m} + U(z) + \frac{1}{2}g_{bb}|\Psi_i(z)|^2 + C \right] \Psi_{i+1}(z).$$
$$(2.32)$$

For a numerical illustration we have chosen to adopt typical system parameters used in experiments at LENS [120] for a condensed gas of ^{87}Rb atoms, setting $\mathcal{N}_b = 10^5$, $N = 200$, and $U_0 = 3.5E_r$ with $E_r = \hbar^2 k_L^2/2m$ being the boson recoil energy. We have also set $2\pi/k_L = 763\,\text{nm}$ and $\delta^2 = 10^{-2}$. Our results are shown in Figure 2.14 where we have plotted the condensate transmittivity through a single-well, a double-well, and a Fibonacci array as a function of T_B/τ_t, $T_B = h/(2Fd)$ being the Bloch oscillation period for the bosons in the double-well lattice and τ_t a time scale to be defined in the following.

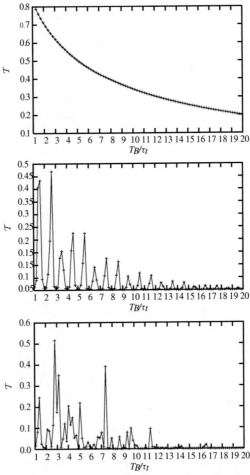

Figure 2.14. Condensate transmittivity through a single-well array (top), a double-well array (center), and a Fibonacci chain (bottom), as a function of T_B/τ_t. The T_B and time scale τ_t in all cases are taken as the Bloch oscillation period and the tunneling time in the double-well lattice.

The condensate transmittivity in the single-well lattice is monotonic and increases with the external force (see top-left panel in Figure 2.14), as expected for a quasi-particle executing Bloch oscillations in a single band and as observed in the experiments of Anderson and Kasevich [96]. In contrast, for a double-well array the center panel of Figure 2.14 shows a very prominent structure of peaks and troughs in the transmittivity. This pattern arises from the interference occurring at $p = -h/4d$ (the left-hand edge of the Brillouin zone in the split-band configuration shown in Figure 2.15) between the condensed atoms that are Bragg scattered at $p = h/4d$ in the lower sub-band and those that have tunnelled into the upper sub-band, travelled through it, and tunnelled back into the lower sub-band at $p = -h/4d$. The time parameters that govern the interference pattern are the Bloch oscillation period $T_B = h/(2Fd)$ and the time τ_t taken by the condensate in coherently tunnelling twice across the minigap ΔE (see Figure 2.15), which is proportional to $\hbar/\Delta E$ [121]. We expect constructive interference in the reflectivity at $p = 0$ in the upper sub-band, $i.e.$ destructive interference in the transmittivity, whenever the ratio T_B/τ_t is an integer number. From the results in Figure 2.14 we find the tunnelling time as $\tau_t = (3\pi^2/8)(\hbar\mathcal{N}_b/N\,\Delta E)$. Of course this specific value of the proportionality constant in the tunnelling time, which is consistent with the Heisenberg principle, depends on the model Hamiltonian that we have assumed. In Appendix A we illustrate the equivalence

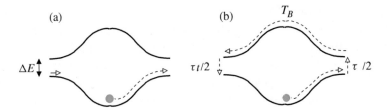

Figure 2.15. The two back-scattering paths governing the interference pattern in the center panel of Figure 2.14. (a) A boson wavepacket travels through the lower sub-band and is Bragg-scattered at $p = h/4d$ to $p = -h/4d$. (b) A boson wavepacket tunnels at $p = h/4d$ into the upper sub-band, travels through it, and tunnels back into the lower sub-band at $p = -h/4d$.

between a condensate wavepacket propagating in a double-well lattice connected to leads and beam-splitting for a light beam travelling through a five-layer optical medium.

Finally, turning to the Fibonacci array in the bottom panel of Figure 2.14, the low values of the transmission coefficient directly reflect low values of the particle current due to scattering against the quasi-periodic disorder. The minima correspond essentially to localization of the atoms induced by the large number of pseudo-gaps in the DOS.

2.4.5. Numerical results for a gas of spin-polarized fermions

We present in this section original results for matter transport in a spin-polarized gas of fermions ^{40}K atoms moving under a constant applied force through optical arrays superposed onto an elongated cigar-shaped harmonic trap. As noted in Section 2.2, the observation of long-lasting Bloch oscillation has been very recently reported for this system in a single-well array [101]. We consider again three types of array at half filling, corresponding to $N_f = 100$ fermions in $N = 200$ sites.

The expression "spin-polarized gas" is in this context a standard shorthand notation to indicate that the fermionic atoms occupy a single Zeeman sublevel in a magnetic trap. The antisymmetry of the many-body wavefunction under exchange has important consequences for the physical behavior of the gas (see for instance [1] for an exhaustive discussion).

The spin-polarized fermions can to a very good approximation be considered as noninteracting, since the Pauli principle inhibits close encounter between them and this suppresses s-wave collisions, which at essentially zero temperature the two-body collisions in higher partial-wave state are weaker by several order of magnitude than typical s-wave collisions between bosonic atoms.

The role of the Pauli principle in keeping apart fermions with identical spins nevertheless results in what is known as the Fermi pressure, which acts as an effective repulsion even between noninteracting fermions. Indeed, a precise mapping exists in 1D between a spin-polarized Fermi gas and a gas of impenetrable bosons in the so-called Tonks limit of infinity strong contact repulsions. In particular, in the ground state of the gas the spin-polarized fermions fill the single-particle energy levels of the confining potential up to a maximum fixed by their number.

In the specific case of present interest, in a single-well lattice at half filling the fermions occupy all the single-particle states in the bottom half of the lowest energy band.

This situation should be contrasted with the case of a bosonic condensate shown in the left side of Figure 15, where the gas in its ground state is all condensed in the zero-momentum single-particle state. While the condensate is driven by an external force through the states of the band structure as if it were a single quasi-particle, coherent motion of the Fermi gas under an imposed bias can still occur if each fermions is in turn accelerated into successively higher band states.

According to the above discussion, the 1D fermion wavefunctions in the $i - th$ well have the same form as in equation (2.29) with $\Psi_i(0) = 1$, and the parameters in the effective 1D Hamiltonian have the same form as in equations (2.30) and (2.32) with the coupling strength g_{bb} set to zero. For a numerical illustration we have chosen the same values of the system parameters as for a condensate in Section 2.4.4, *i.e.* $U_0 = 3.5E_r$, $2\pi/k_L = 763$ nm, and $\delta^2 = 10^{-2}$. Our results are shown in Figure 2.16, where we have plotted the transmittivity of the Fermi gas through a single-well, a double-well, a Fibonacci array, and a randomly distributed sequence as a function of T_B/τ_t, with $T_B = h/(2Fd)$ being the Bloch period in the double-well lattice and τ_t a suitable time-scale parameter (see below).

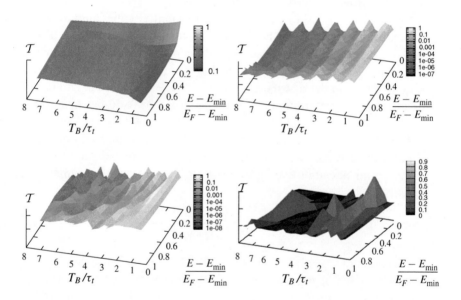

Figure 2.16. Transmittivity $\mathcal{T}$ of a spin-polarized Fermi gas through a single-well (top-left), a double-well (center), a Fibonacci array (bottom-left), and a randomly distributed sequence (bottom-right), as a function of T_B/τ_t and of $(E - E_{min})/(E_F - E_{min})$. The time scale T_B and τ_t in all cases are taken as the Bloch oscillation period and the tunneling time in the double-well array.

It is seen from Figure 2.16 that fermions at the Fermi level in the single-well array have a better coupling with the outgoing lead at large values of T_B, namely for low values of the drive, while those at very low energies have a greater coefficient of transmittivity at large values of the drive.

The behavior of the fermion transmittivity through a double-well array is illustrated in the center panel of Figure 2.16. As in the boson case, we expect destructive interference in the transmission whenever the ratio between the Bloch period and the time to tunnel twice trough the minigap ΔE is an integer number. This sets the value of the fermion tunnelling time at $\tau_t = 6\pi\hbar/\Delta E$.

The fermions travelling through the Fibonacci chain explore an energy spectrum which on average is rather more akin to the single-well band structure than to the doubled-well one (see Figure 2.16). Nevertheless, the transmittivity shows marked peaks and troughs as a consequence of quasi-periodic disorder (see the bottom-left panel in Figure 2.16).The positions of the peaks do not show any regularity and depend not only on the applied drive but also on the fermion energy.

As a last case, we show the result for transmittivity in a randomly distributed sequence (see the bottom-right panel in Figure 2.16). In this case, two types of wells are arranged completely at random sequence which has been produced from a random number generator. The effect of randomness in the distribution, is to strongly diminish the contrast in the interference pattern and the positions of the peak do not posses any regular pattern.

With the aim of a direct comparison with the transmittivity of Bose-Einstein condensate, we have plotted in Figure 2.17 the transmittivity of the Fermi gas after averaging over all incoming states, namely the quantity

$$\bar{T} = \frac{\int_{E_{min}}^{E_F} T(E)\, D(E)\, dE}{\int_{E_{min}}^{E_F} D(E)\, dE} \tag{2.33}$$

as a function of T_B/τ, $D(E)$ being the DOS of the incoming lead.

The behavior of the averaged fermion transmittivity is qualitatively similar to that shown in Figure 2.14 for bosons. In particular, Bloch oscillations with a period $h/(Fd)$ are executed by the whole distribution of fermions once the coupling to the outgoing lead is cut. However, inter-sub-band coherence is strongly reduced and the consequences of quasi-periodic disorder are stronger for the Fermi gas. In the case of completely random sequence, the averaged transmittivity reaches to the minimum value respect to the other cases.

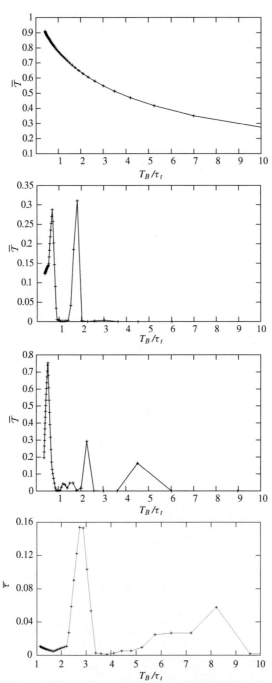

Figure 2.17. Averaged transmittivity $\bar{\mathcal{T}}$ of a spin-polarized Fermi gas through a single-well (top), a double-well (center up), a Fibonacci array (center-down), and a randomly distributed sequence (bottom), as a function of T_B/τ_t.

Chapter 3
Bethe-ansatz density-functional study of Fermi gases in one-dimensional optical lattices

This Chapter contains our analyses of the ground state of a two-component Fermi gas with repulsive interactions, subjected to an external confining potential inside a 1D lattice. Our calculations use novel versions of Density Functional Theory (DFT) and Spin-Density Functional Theory (SDFT). After a brief introduction of these theories, we switch to the lattice version of DFT- the so-called Site-Occupation Functional Theory (SOFT) that was introduced by Gunnarsson and Schönhammer to study the band-gap problem in the context of *ab inito* theories of fundamental energy band gaps for electrons in semiconductors and insulators. The reference system for our application of SOFT is the single-band Hubbard model, which in the homogeneous case has been solved exactly by Lieb and Wu [122]. The exchange-correlation energy of the Lieb-Wu solution can be used to construct a local spin-density approximation based on a Bethe-Ansatz solution for the ground state of the trapped gas. We first review some recent work by Gao Xianlong *et al.* [Phys. Rev. B **73**, 165120 (2006)] on unpolarized gases, highlighting in particular certain ground-state phases which exhibit phase separation between metallic and Mott-insulating phases. By generalizing this work we present detailed numerical results for the spin-resolved atom-density profiles obtained from Kohn-Sham spin-density-functional calculations. We investigate both spin-independent and spin-dependent parabolic external potentials. In the latter case we find that phase separation of the two spin populations occurs in a certain region of parameter.

3.1. Density functional theory at zero temperature

A most important development of electron-gas theory has been the discovery of the DFT approach [123, 124, 125] to study the electronic structure of atoms, molecules, liquids and solids (inhomogeneous electronic systems in which electrons move in the field of nuclei or of ionic cores). The method combines at a formally exact level the idea of the Thomas-

Fermi method (the electronic structure is described through its density distribution $n(\mathbf{r})$) and the idea of the Hartree-Fock method (the electronic structure is described through self-consistently determined single-electron orbitals). There already are available in the literature a number of books on the foundations and the applications of DFT [41, 126]: our aim here will simply be to introduce the Hohenberg-Kohn (HK) theorem and Kohn-Sham (KS) scheme.

3.1.1. The Hohenberg-Kohn theorem

Consider an interacting non-relativistic many-fermion system in an external static scalar potential $V_{\text{ext}}(\mathbf{r})$ with an Hamiltonian

$$
\begin{aligned}
\hat{\mathcal{H}} &= \hat{T} + \hat{W} + \hat{V}_{\text{ext}} \\
&= -\frac{\hbar^2}{2m} \sum_{\sigma} \int d\mathbf{r}\, \hat{\psi}_\sigma^\dagger(\mathbf{r}) \nabla^2 \hat{\psi}_\sigma(\mathbf{r}) \\
&\quad + \frac{1}{2} \sum_{\sigma,\sigma'} \int d\mathbf{r} \int d\mathbf{r}'\, \hat{\psi}_\sigma^\dagger(\mathbf{r}) \hat{\psi}_{\sigma'}^\dagger(\mathbf{r}')\, w(\mathbf{r},\mathbf{r}')\, \hat{\psi}_{\sigma'}(\mathbf{r}') \hat{\psi}_\sigma(\mathbf{r}) \\
&\quad + \sum_{\sigma} \int d\mathbf{r}\, \hat{\psi}_\sigma^\dagger(\mathbf{r}) V_{\text{ext}}(\mathbf{r}) \hat{\psi}_\sigma(\mathbf{r}),
\end{aligned}
\tag{3.1}
$$

where $w(\mathbf{r},\mathbf{r}')$ is a generic pair-potential and $\hat{\psi}_\sigma^\dagger(\mathbf{r})$ ($\hat{\psi}_\sigma(\mathbf{r})$) is the Schrödinger field operator which creates (annihilates) a particle in position $\mathbf{r}$ with spin σ. Suppose to subject the system to different external potentials. Let v be a set of local one-particle potentials with the property that the solution of each eigenvalue problem

$$
\hat{\mathcal{H}}\,|\varphi\rangle = (\hat{T} + \hat{W} + \hat{V}_{\text{ext}})\,|\varphi\rangle = E\,|\varphi\rangle, \quad \hat{V}_{\text{ext}} \in v
\tag{3.2}
$$

leads to a non-degenerate ground state (GS) for a system of N fermions:

$$
\hat{\mathcal{H}}\,|\Psi\rangle = E_{\text{GS}}|\Psi\rangle.
\tag{3.3}
$$

Solutions of the Schrödinger equation define a map C between the set of one-particle potentials v and the set of ground-state Ψ:

$$
C : v \ni \hat{V}_{\text{ext}} \longrightarrow |\Psi\rangle \in \Psi.
\tag{3.4}
$$

In this map Ψ contains no element which is not associated with some element of v.

Next, for all ground-state wavefunctions contained in Ψ, it is possible to calculate the ground state densities

$$
n(\mathbf{r}) = \langle \Psi |\, \sum_{\sigma} \hat{\psi}_\sigma^\dagger(\mathbf{r})\, \hat{\psi}_\sigma(\mathbf{r})\, |\Psi\rangle,
\tag{3.5}
$$

establishing a second map

$$D : \Psi \ni |\Psi\rangle \longrightarrow n(\mathbf{r}) \in \mathcal{N}. \tag{3.6}$$

It is possible to prove that the maps C and D are also one-to-one and thus fully invertible. The proof of being one-to-one for the map C is simple and is based on the fact that the external potential is a multiplicative operator. The same proof for the map D uses the variational Rayleigh-Ritz principle on the ground-state energy [126].

First statement of the Hohenberg-Kohn theorem

The ground state expectation value of **any** *observable $\hat{O}$ is a* **unique** *functional of the exact ground state density:*

$$\langle \Psi[n]|\hat{O}|\Psi[n]\rangle = \mathcal{O}[n]. \tag{3.7}$$

This follows immediately from the fact that unique inversion of the map D is possible:

$$D^{-1} : \mathcal{N} \ni n(\mathbf{r}) \longrightarrow |\Psi[n]\rangle \in \Psi. \tag{3.8}$$

The physical implication of such a statement is profound: in principle the calculation of any observable needs the knowledge of the many-body wavefunction which is a function of $2N$-variables. The previous statement says that the density $n(\mathbf{r})$ is instead sufficient. Moreover, the full inverse map $(CD)^{-1}$ tells that knowledge of the ground state density determines the external potential (to within a trivial constant), and thus, as the kinetic energy and the interparticle potential are specified, the full Hamiltonian.

Second statement of the Hohenberg-Kohn theorem

The exact ground state density can be determined by minimization of the functional

$$E_{V_0}[n] \equiv \langle \Psi[n]|\hat{T} + \hat{W} + \hat{V}_0|\Psi[n]\rangle, \tag{3.9}$$

where $\hat{V}_0$ is the external potential of a system with ground state density $n_0(\mathbf{r})$ and ground state energy E_0. In short

$$E_0 = \min_{n \in \mathcal{N}} E_{V_0}[n]. \tag{3.10}$$

Notice that the map D^{-1} does not depend on the external potential of the particular system under consideration. Thus, writing

$$E_{V_0}[n] = F_{\text{HK}} + \int d\mathbf{r}\, V_0(\mathbf{r})\, n(\mathbf{r}), \tag{3.11}$$

with

$$F_{HK} \equiv \langle \Psi[n] \,|\, \hat{T} + \hat{W} \,|\, \Psi[n] \rangle \,, \qquad (3.12)$$

it is possible to state that the functional F_{HK} is *universal*, *i.e.* independent of $\hat{V}_0$.

3.1.2. The Kohn-Sham scheme

The variational theorem of Hohenberg and Kohn allows the determination of the exact ground state density of a specified many-body system. Possible advantages of a replacement of the direct variation with respect to the density by the intermediary of an orbital picture were first emphasized by Kohn ans Sham. Interestingly, this approach owes its success and popularity partly to the fact that it does not exclusively work in terms of the particle (or charge) density, but brings a special kind of wave functions (single-particle orbitals) back into the problem. As a consequence DFT then looks formally like a single-particle theory, although many-body effects are still included *via* the so-called *exchange-correlation* (xc) functional. We will now see in some detail how this is done.

Consider an auxiliary system of N noninteracting particles described by the Hamiltonian

$$\hat{\mathcal{H}}_s = \hat{T} + \hat{V}_s \,. \qquad (3.13)$$

According to the Hohenberg-Kohn theorem, there exists a unique energy functional

$$E_s[n] = T_s[n] + \int d\mathbf{r} \, V_s(\mathbf{r}) \, n(\mathbf{r}), \qquad (3.14)$$

for which the variational equation

$$\frac{\delta E_s[n]}{\delta n(\mathbf{r})} = 0 \,, \qquad (3.15)$$

yields the exact ground state density $n_s(\mathbf{r})$ corresponding to $V_s(\mathbf{r})$. $\hat{T}_s$ denotes the universal kinetic energy functional of *non-interacting* particles. The central assertion used in establishing the Kohn-Sham scheme is: for any interacting system, there exists a local single-particle potential $V_s(\mathbf{r})$ such that the exact ground state density of the interacting system equals the ground state density of the auxiliary problem,

$$n(\mathbf{r}) = n_s(\mathbf{r}). \qquad (3.16)$$

If the ground state of $\hat{\mathcal{H}}_s$ is nondegenerate, the ground state density $n_s(\mathbf{r})$ [and thus $n(\mathbf{r})$] possesses a unique representation

$$n(\mathbf{r}) = \sum_{i=1}^{N} |\varphi_i(\mathbf{r})|^2 \qquad (3.17)$$

in terms of the lowest N single-particle orbitals obtained from the Schrödinger equation:

$$\left[-\frac{\hbar^2}{2m}\nabla^2 + V_{KS}(\mathbf{r})\right]\varphi_i(\mathbf{r}) = \varepsilon_i\,\varphi_i(\mathbf{r}),\quad \varepsilon_1 \leq \varepsilon_2 \leq \ldots \quad (3.18)$$

The case of a degenerate level ε_N is discussed in detail in Ref. [126].

Once the existence of a potential $V_{KS}(\mathbf{r})$ generating a given density $n(\mathbf{r})$ via equations (3.17) and (3.18) is assumed, uniqueness follows from the Hohenberg-Kohn theorem. Thus the single-particle orbitals in equation (3.18) are unique functionals of the density $n(\mathbf{r})$,

$$\varphi_i(\mathbf{r}) = \varphi_i[n](\mathbf{r})\,, \quad (3.19)$$

and the non-interacting kinetic energy

$$T_s[n] = \sum_{i=1}^{N}\int \varphi_i^*(\mathbf{r})\left(-\frac{\hbar^2}{2m}\nabla^2\right)\varphi_i(\mathbf{r})\,d\mathbf{r} \quad (3.20)$$

is a unique functional of $n(\mathbf{r})$ as well.

Now considering a particular interacting system with external potential $V_{ext}(\mathbf{r})$ and the ground state density $n_0(\mathbf{r})$, the single-particle Kohn-Sham potential which generates $n_0(\mathbf{r})$ via equations (3.17) and (3.18) is [126]:

$$V_{KS}(\mathbf{r}) = V_{ext}(\mathbf{r}) + V_H(\mathbf{r}) + V_{xc}([n_0];\mathbf{r})\,, \quad (3.21)$$

where for the pair-potential $w(\mathbf{r},\mathbf{r}')$, the Hartree potential $V_H(\mathbf{r})$ is given by

$$V_H(\mathbf{r}) = \int d\mathbf{r}'\,w(\mathbf{r},\mathbf{r}')\,n_0(\mathbf{r}'), \quad (3.22)$$

while the exchange-correlation potential $V_{xc}([n_0];\mathbf{r})$ is given by:

$$V_{xc}[n_0](\mathbf{r}) \equiv \left.\frac{\delta E_{xc}[n]}{\delta n(\mathbf{r})}\right|_{n_0(\mathbf{r})}. \quad (3.23)$$

Here, the exchange-correlation energy functional $E_{xc}[n]$ is formally defined by:

$$E_{xc}[n] \equiv F_{HK}[n] - \frac{1}{2}\int d\mathbf{r}\int d\mathbf{r}'n(\mathbf{r})w(\mathbf{r},\mathbf{r}')n(\mathbf{r}') - T_s[n]. \quad (3.24)$$

Therefore the effective one-body potential in equation (3.21) is the sum of the external potential, the classical Hartree potential, and the exchange-correlation potential. It must be determined self-consistently with the equilibrium particle density profile.

The previous definition comes from a special decomposition (usually called Kohn-Sham decomposition) of the Hohenberg-Kohn ground state energy functional given in equation (3.11):

$$E_{V_0}^{\mathrm{KS}}[n] = T_s[n] + \int d\mathbf{r}\, V_{\mathrm{ext}}(\mathbf{r})\, n(\mathbf{r}) + E_{\mathrm{H}}[n] + E_{\mathrm{xc}}[n], \qquad (3.25)$$

the Hartree contribution being defined by

$$E_{\mathrm{H}}[n] \equiv \frac{1}{2} \int d\mathbf{r} \int d\mathbf{r}'\, n(\mathbf{r})\, w(\mathbf{r}, \mathbf{r}')\, n(\mathbf{r}'). \qquad (3.26)$$

equations (3.17) and (3.18) with the potential (3.21) represent the classic Kohn-Sham scheme. Since the single-particle potential depends on the density, the whole set of equations has to be solved in a self-consistent manner. Once one has a converged solution n_0, one can calculate the *total* energy from equation (3.25) or, equivalently and more conveniently, from [126]

$$E_0 = \sum_i^N \varepsilon_i - \frac{1}{2} \int d\mathbf{r} \int d\mathbf{r}'\, n_0(\mathbf{r})\, \omega(\mathbf{r}, \mathbf{r}')\, n_0(\mathbf{r}')$$

$$- \int d\mathbf{r}\, V_{\mathrm{xc}}\, n_0(\mathbf{r}) + E_{\mathrm{xc}}[n_0]. \qquad (3.27)$$

Equation (3.27) shows that E_0 is not simply the sum of all ε_i [127]. In fact it should be emphasized that the ε_i are introduced as artificial objects: they are the eigenvalues of an auxiliary single-body Schrödinger equation whose eigenfunctions (orbitals) yield the correct density. It is only this density that has physical meaning in the Kohn-Sham equations. The Kohn-Sham eigenvalues, on the other hand, bear only a semiquantitative resemblance with the true energy spectrum [128], but are not to be trusted quantitatively.

3.1.3. Local-density approximation

So far DFT has been presented as a formal mathematical framework for viewing electronic structure from the perspective of the electron density $n(\mathbf{r})$. This mathematical framework has been motivated by physical considerations, but to make concrete use of it we require effective approximations for $F_{\mathrm{HK}}[n(\mathbf{r})]$ in the HK formulation, and for $E_{\mathrm{xc}}[n(\mathbf{r})]$ in the KS formulation. These approximations reflect the physics of electronic structure and come from outside of DFT [127].

The simplest, and at the same time remarkably serviceable, approximation for $E_{\mathrm{xc}}[n(\mathbf{r})]$ is the so-called local-density approximation (LDA). To fermions: in particular two fermions with half-integer spin in a two-body bound state will produce an integer-spin composite boson. In recent experiments the fermionic particles are atoms, forming a diatomic molecule in a two-body bound state [66, 67, 68]. Below a critical temperature an ensemble of these diatomic molecules forms a BEC [69, 70, 71]. The left side of Figure 1.22 represents a superfluid containing these pairs. Two different spin states are required if the fermions are to pair *via* s-wave ($l = 0$) interactions.

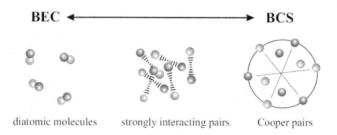

Figure 1.22. Cartoon illustration of the continuum of pairing in the BCS-BEC crossover. From [72].

In the Cooper pairing mechanism instead fermion pairing occurs in momentum space. Cooper considered two fermions with equal and opposite momenta on the Ferm surface inside a Fermi sea [73]. The energy of the two fermions is found to be less than the value of $2E_F$ for arbitrarily weak attractive interactions. This result is in contrast to that for two fermions in vacuum: in this case there will be no bound state unless the attraction reaches a threshold. The key difference between the two situations arises from Pauli blocking, which in the Cooper-pair case prevents the two fermions from occupying momentum states below the Fermi momentum [73]. The BCS state is a superfluid condensate of such Cooper pairs (Figure 1.22 right side). In space the pairs are highly overlapping and cannot be simply considered as composite bosons. In the BCS limit the momentum distribution only changes from the usual Fermi sea in an exponentially small region around the Fermi surface.

It is interesting to consider what happens if diatomic molecules become more and more weakly bound, to the point where the binding energy E_b of the molecules becomes less than the Fermi energy E_F. One could also consider increasing the interaction energy of a Cooper paired state until it is close to E_F. The essence of the BCS-BEC crossover is that these two viewpoints describe the same physical state. As the interaction between fermions is increased there will be a continuous change, or *crossover*, between a BCS state of Cooper pairs and a BEC state of diatomic molecules. The point where two fermions in vacuum would have zero binding energy is considered the cusp of the crossover problem, and pairing in such a state is represented in the middle of Figure 1.22: many-body effects are required for the pairing as in the BCS state, but there is some amount of spatial correlation as with diatomic molecules. The pair size is on the order of the spacing between fermions, and the system is strongly interacting.

The effective strength of the interactions in dilute atomic gases can be modified in several ways. In addition to varying the density of the sample or to localizing the atoms on a lattice [1], one may exploit a Feshbach resonance. This occurs when the collision energy of two free atoms is the same as that of a quasibound molecular state and on approaching it the scattering length can increase by some orders of magnitude and also change sign. The high-dilution condition breaks down as the gas is driven close to a Feshbach resonance.

At a magnetic-field Feshbach resonance, relatively small changes in magnetic field strength can have dramatic effects on the effective interactions in an ultracold gas. On one side of the Feshbach resonance value, for example at a slightly higher field strength in the case of experiments with ^{40}K atoms, the atoms have very strong effectively attractive interactions (see Figure 1.23).

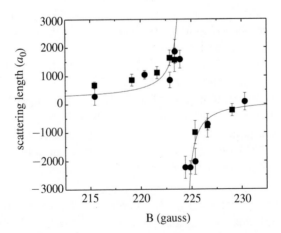

Figure 1.23. Scattering length for ^{40}K atoms in two different Zeeman sublevels driven across a Feshbach resonance by varying the magnetic field B, as measured by Regal and Jin [74].

Observation of a Bose-Einstein condensate from a Fermi gas

In 2003, Jin's group prepared a ^{40}K molecular BEC in a single focused laser trap by lowering the trap power and obtained 2.4×10^4 molecules of ^{40}K [70]. In this experiment Greiner *et al.* with a relatively slow sweep of an applied magnetic field that converts most of the fermionic atoms into bosonic molecules and an initial atomic gas below $T/T_F = 0.17$, observed a molecular condensate in time-of-flight absorption images taken immediately following the magnetic-field sweep.

Figure 1.24 shows a bimodal distribution for an experiment starting with an initial temperature of $0.19\,T_F$ ($0.06\,T_F$) for the left (right) graph which indicates that the cloud of weakly bound ^{40}K molecules experienced a phase transition to BEC with an temperature below the critical temperature T_c. The fits (lines) yield no condensate fraction and a temperature of of $T = 0.90 T_c$ for the left graph, and a 12% condensate fraction and a temperature of the thermal component of $T = 0.49 T_c$ for the right graph. Here, T_c is the calculated critical temperature for a non-interacting BEC in thermal equilibrium.

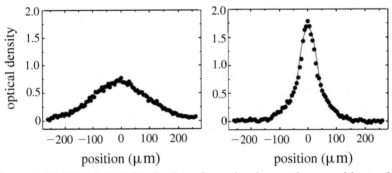

Figure 1.24. Momentum distribution of a molecule sample created by applying a magnetic-field ramp to an atomic Fermi gas with an initial temperature of $0.19\,T_F$ ($0.06\,T_F$) for the left (right) picture. In the right sample the molecules form a Bose-Einstein condensate. The lines illustrate the result of bimodal fits. From [70].

To obtain the condensate information of the molecular cloud, Greiner *et al.* [70] studied dependence of the molecular condensate fraction N_0/N on the scaled temperature T/T_c or the magnetic field ramp rate or the initial scaled temperature T/T_F of the Fermi gas. They found that:

(1) with the reduction of the scaled temperature T/T_c starting from 0.8, which was realized by lowering the trap laser power, the molecular

condensate fraction N_0/N would be quickly increased from zero to the maximum of $\sim 16\%$, and an actual critical temperature was $0.8T_c$. This decrease is due to the repulsive interactions in a trapped molecular gas;

(2) the largest molecular condensate fraction N_0/N was obtained within the B-field ramp time of 3-10 ms, which shows that the creation of a molecular BEC required an efficiently slow B-field tuning;

(3) a molecular BEC was formed when the initial temperature was below $0.17T/T_F$, and with the dropping of the initial scaled temperature T/T_F of the Fermi gas, the molecular condensate fraction N_0/N would increase rapidly and reached the maximum value at $T/T_F = 0.05$.

It should be noted that there are two different routes to create a molecular BEC. The first is a straightforward approach where ultracold molecules are generated in a degenerate Fermi gas and then evaporatively cooled to a temperature below the critical temperature for the BEC phase transition by lowering the trap laser power, which was done by Grimm's group [71]. The second is a very different approach, in which the interaction in a deep degenerate Fermi gas of atoms is tuned to be more and more attractive until a molecular BEC are created as demonstrated by Jin's group [70]. Here, there was no direct cooling of the molecules, a Feshbach resonance was used to adiabatically change the interaction strength in the original Fermi atomic gas until a molecular condensate is formed. The realization of these molecular BEC will provide an ideal experimental platform for studies of the BCS-BEC crossover, including the formation of Fermi atomic pairs (*i.e.*, Cooper pairs) and its condensate as well as BCS-type superfluidity [75].

Condensation of Fermi atom pairs: fermionic condensates

In 2004, Jin's group observed condensation of Fermi-atom pairs near and on both sides of the Feshbach resonance [76], which corresponds to the BCS-BEC crossover regime. The observed condensation was on the BCS ($a < 0$) side of the Feshbach resonance. Here the two-body physics of the resonance no longer supports the weakly bound molecular state. So only the cooperative many-body effects can result in this condensation of Fermi atom pairs [77].

Starting from a ^{40}K molecular BEC in an optical dipole trap, Regal *et al.* explored the behavior of the sample when the values of applied magnetic field B_{hold} was swept slowly on either of the Feshbach resonance $B_0 = 202.1 G$ and realized condensation of fermionic atom pairs on the BCS side of the resonance [76]. The distance between the B_{hold}

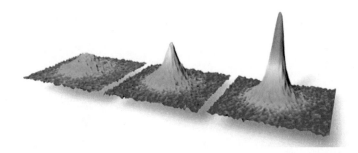

Figure 1.25. Time-of-flight images showing condensation of fermionic atom pairs. The images, taken after the projection of the fermionic system onto a molecule gas, are shown for $\Delta B = 0.12$, 0.25, and 0.55 G (right to left) on the BCS side of the resonance. Here, ΔB indicates the distance from a Feshbach resonance. The original atom gas starts at $(T/T_F)^0 = 0.07$. The figure has been taken from the JILA webpage.

and B_0 is denoted as $\Delta B = B_{\text{hold}} - B_0$. Figure 1.25 presents the sample time-of-flight absorption images for the fermionic condensate, in which the maximum condensed number was 1.0×10^4 Fermi atom pairs. In addition, Regal *et al.* studied the dependence of condensate fraction on both T/T_F and the inverse sweep speed, and found that the smaller the T/T_F is, the higher the N_0/N is, while the condensate on the BCS side of the resonance does not depend on the sweep speed. As in the case of BEC, the observed resonant fermionic condensate may show the predicted BCS-based superfluidity.

Also Bose-Einstein condensation of pairs of fermionic atoms in an ultracold ^{6}Li gas at magnetic fields above a Feshbach resonance was observed in Ketterle's group [78]. They accurately determined the position of the resonance to be 822 ± 3 G. They also studied the dependence of the pair condensate fraction on both the magnetic field and the temperature, and found that the observed pair condensation appeared in the regime where $T/T_F < 0.2$ and $k_F|a| > 1$.

Chapter 2
Theory of matter transport
in quasi-1D arrays

In this chapter, we present the results that we have obtained for the coherent transport by electrons through quantum-dot arrays [79] and by
quantum gases through equi-spaced potential wells [80, 81]. We use a
Green's function approach and a renormalization method in order to calculate Density-of-States for various array. Based on this approach, we
obtain also transmittivity and Density-of-states in the presence of external leads. The final part of the chapter deals with transport of matter by
a Bose-Einstein condensate and by a spin-polarized Fermi gas through a
linear array of equi-spaced potential wells. We consider different types
of array, namely a periodic arrangement of either identical wells or pairs
of different wells, a quasi-periodic sequence , and a randomly-distributed
sequence.

2.1. Introduction

The electronic transport properties of linear arrays of scatterers under
an applied electric field have been extensively studied both for their intrinsic theoretical interest and for their relevance to applications. The
systems of current interest include polymers and molecular wires, nan-
otubes, quantum wires, and arrays of quantum dots. A basic viewpoint
on electron transport was put forward in a seminal paper by Landauer
[82], who showed that the dc conductance is proportional to the electron
transmission coefficient through the array. From this viewpoint Landauer
proceeded to show in a simple and direct manner that the resistivity of a
completely disordered 1D lattice grows exponentially rather than linearly
with the number of scatterers, this result being related to localization of
the wave function with prevalence of exponential decay. Within a tight-
binding framework the site energies in the array can be taken to vary
according to a chosen law (such as periodic, quasi-periodic, or random)
and the density of electron states and the transmission coefficient can be
calculated by Green's functions methods (see for instance [83]).

Confinement of charge carriers in a semiconductor quantum dot can be achieved by electrical gating of a 2D gas of carriers in combination with etching of the material [84, 85], and an electrically controlled array of quantum dots can be realized by electrodes confining single carriers to the dot regions, with the tunnel between neighboring dots being controlled by gating. The tight-binding model has been used to treat transport through quantum dots coupled to carrier reservoirs [86], Coulomb effects on transmittance [87], magneto-conductance in chaotic arrays [88], Kondo resonances and Fano anti-resonances in transport [89], and transport through quantum-dot networks [90, 91, 92]. A simplifying assumption treats a quantum dot as a potential well, thus omitting a detailed account of its internal structure but allowing analytical results to be obtained for linear arrays (see for instance [93, 94]).

Advances in the preparation and manipulation of ultra-cold atomic gases have very significantly broadened in recent years the scope of the study of phase-coherent quantum transport through arrays of potential wells (for a recent review see [1]). A standing electromagnetic wave resulting from the interference of two counter-propagating laser beams, which are detuned away from an atomic absorption line, can be superposed onto a highly elongated magnetic trap to provide an almost ideal linear periodic array of potential wells for an atomic gas. Experiments on bosonic gases in such an "optical lattice" have reported the observation of Bloch oscillations both for ultra-cold atoms [95] and for a Bose-Einstein condensate [96], Landau-Zener tunnelling [97], Josephson-like oscillations [98, 99], a superfluid-to-Mott-insulator transition [60], and the 1D band structure [100]. Collective Bloch oscillations lasting for very many periods have been observed in a spin-polarized Fermi gas inside an optical lattice, opening a novel road to high-precision interferometry and to the measurement of forces with microscopic spatial resolution [101]. The ideal conditions of extremely low temperature and extremely high purity in which experiments on atomic gases can be carried out offer a virtually unique opportunity for the investigation and testing of transport theory at unprecedented levels of accuracy and depth.

In this chapter we aim to present in parallel the essential elements of the theory of coherent transport of electrons and of atomic Bose or Fermi gases under a constant applied force through a linear array of equi-spaced potential wells described by a tight-binding Hamiltonian. We consider three main types of array, namely periodic arrangements of either identical wells or pairs of different wells, and a quasi-periodic sequence of pairs of different wells arranged according to the Fibonacci series. In Section 2.2 we introduce these arrays and describe their main features in a tight-binding viewpoint, namely the density of single-particle states

(DOS) and the reduction of a linear array to an effective dimer by deci-mation/renormalization of the sites [102, 103]. In Section 2.3 we review dc transport by Fermi-level electrons through a linear array of scatterers connected to external leads, with main attention on how an increasing number of scatterers modifies the role of the leads and the transport prop-erties of the array [79]. We explicitly show how, as the length of the array increases, localization occurs whenever the Fermi level lies in a gap of the density of states, arising either from period doubling or from quasi-periodic disorder. In Section 2.4 we turn to atomic gases, starting with a brief description of how the various types of array can be created from the interference of suitably arranged laser beams.

We introduce the local density of states as a useful concept for the calculation of the transmission coefficient of a gas through an extended array, before proceeding to separately discuss transport of matter by a Bose-Einstein condensate and by a spin-polarized Fermi gas. A conden-sate that has been adiabatically loaded into an optical lattice keeps phase coherence between its fragments in different sites, provided that the lat-tice barrier is not too high and still allows tunnel between neighboring sites. In this situation a weak external force accelerates the condensate through the band states of the lattice as if it were a single quasi-particle [104, 105]. The main phenomena of interest here are:

(i) the Bloch oscillations of the condensate and their mapping into cur-rent flow across a Josephson junction [106, 107];
(ii) the interference patterns between Bragg scattered and inter-subband tunneling wave packets that arise on period doubling [108]; and
(iii) the localization of matter waves induced by the introduction of quasi-periodic disorder [109].

The interference patterns displayed by coherent matter waves propagat-ing through linear arrays have a precise equivalent in beam-splitting ex-periments in ordinary optics [110], as we record in Appendix A. In the case of a spin-polarized Fermi gas, on the other hand, in the absence of dissipative scattering the whole distribution is coherently accelerated by an external force through the band states of the array. We conclude Sec-tion 2.4 by presenting original results for transport of an atomic Fermi gas through linear arrays, showing that very similar behaviors to those of a Bose-Einstein condensate are found after evaluating the transmission coefficient of each individual fermion and averaging over the distribution of the incoming fermions.

2.2. The Green's function approach for density-of-states in 1D tight-binding model

In this section, we explain the decimation-renormalization scheme to convert a finite-range 1D lattice to an effective lattice in which its Green's function can be calculated. We use the Kirkman-Pendry relation to evaluate the density-of-states. This relation uses matrix element of Green's function to calculate density-of-states. The numerical results are given in Section 2.2.2.

2.2.1. The decimation-renormalization procedure

We consider 1 1D sequence of N equally spaced potential wells, to be occupied by electrons or atoms. In the tight-binding model the Hamiltonian is written as

$$\mathcal{H}_{\mathrm{w}} = \sum_{i=1}^{N} \{ e_i \, |i\rangle \, \langle i| + (t_{i,i+1} \, |i\rangle \, \langle i+1| + t_{i+1,i} |i+1\rangle \, \langle i|) \}, \quad (2.1)$$

where the site energy e_i corresponds to the lowest energy level in the i-th well and $t_{i,i+1} = t_{i+1,i}$ is the hopping energy between the i-th and the $(i+1)$-th well. In the following we focus on three types of linear array as schematically illustrated in Figure 2.1 . The first is an array of identical wells, while in the second two different kinds of wells (A and B say) alternate. In the third the two kinds of wells are arranged according to the Fibonacci sequence F_n, namely, starting with $F_1 = A$ and $F_2 = B$, the rest of the chain is built with the rule $F_n = F_{n-1} \, F_{n-2}$. For a given separation between adjacent wells, the period of the second array is twice that of the first, while the third array is commonly denoted as quasi-periodic.

The related Green's function at energy E is

$$G(E) = \frac{1}{E - \mathcal{H}_{\mathrm{w}}}. \quad (2.2)$$

Here (and whenever needed in the following) the energy is a complex variable with an infinitesimal positive imaginary part. To evaluate the density-of-states $D(E)$ (DOS) we use the Kirkman-Pendry relation [111], which relates it to the matrix element $G_{1,N}(E)$ expressing coherence between the first and the last site of the chain,

$$D(E) = \frac{1}{\pi} \, \mathrm{Im} \, \frac{\partial}{\partial E} \ln G_{1,N}(E). \quad (2.3)$$

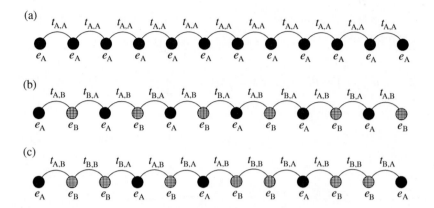

Figure 2.1. Schematic Hamiltonian for (a) an array of identical wells, (b) a regular alternation of two kinds of wells, and (c) a Fibonacci array, for the case $N = 12$.

The matrix element $G_{1,N}(E)$ can in turn be evaluated by reducing the chain to an effective dimer through decimation of the intermediate sites, as is schematically shown in Figure 2.2 (see for instance [112, 113]).

The dimer contains just the first and the last well with renormalized site energies and hopping energy. Its Hamiltonian $\tilde{\mathcal{H}}_w(E)$ depends on the energy E and is expressed as a 2×2 matrix,

$$\tilde{\mathcal{H}}_w(E) = \begin{pmatrix} \tilde{\varepsilon}_1^{(N-2)}(E) & \tilde{t}_{1,N}(E) \\ \tilde{t}_{N,1}(E) & \tilde{\varepsilon}_N^{(N-2)}(E) \end{pmatrix}. \tag{2.4}$$

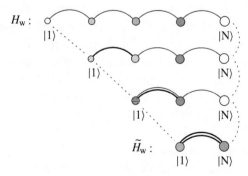

Figure 2.2. Schematic representation of the Hamiltonian $\mathcal{H}_w$ and of the decimation/renormalization procedure leading to the effective Hamiltonian $\tilde{\mathcal{H}}_w$.

Here, the effective site and hopping energies are determined by the recursive relations

$$\tilde{\varepsilon}_1^{(j)}(E) = \tilde{\varepsilon}_1^{(j-1)}(E) + \tilde{t}_{1,j+1}(E) \, \frac{1}{E - \tilde{\varepsilon}_{j+1}^{(j-1)}(E)} \, t_{j+1,j+2}, \quad (2.5)$$

$$\tilde{\varepsilon}_{j+2}^{(j)}(E) = e_{j+2} + t_{j+2,j+1} \, \frac{1}{E - \tilde{\varepsilon}_{j+1}^{(j-1)}(E)} \, \tilde{t}_{j+1,1}(E), \quad (2.6)$$

$$\tilde{t}_{1,j+2}(E) = \tilde{t}_{1,j+1}(E) \, \frac{1}{E - \tilde{\varepsilon}_{j+1}^{(j-1)}(E)} \, t_{j+1,j+2} \quad (2.7)$$

and $\tilde{t}_{j+1,1} = \tilde{t}_{1,j+1}$ for $j \geq 1$, the initial values being given by the original Hamiltonian parameters $\tilde{\varepsilon}_i^{(0)}(E) = e_i$ and $\tilde{t}_{1,2}(E) = t_{1,2}$. By direct inversion of $(E - \tilde{\mathcal{H}}_w(E))$ we obtain

$$G_{1,N}(E) = \frac{\tilde{t}_{N,1}(E)}{[E - \tilde{\varepsilon}_1^{(N-2)}(E)][E - \tilde{\varepsilon}_N^{(N-2)}(E)] - [\tilde{t}_{1,N}(E)]^2}. \quad (2.8)$$

We then use equation (2.3) to calculate the DOS.

2.2.2. Numerical results for Density-of-States

In the illustrative calculations that we report in this Section and in Section 2.4 we have used the values of the parameters $e_A = -0.25$ eV, $e_B = 0.25$ eV, $t_{A,B} = 1.1$ eV and $t_{A,A} = t_{B,B} = 1.0$ eV. In Figure 2.3 and Figure 2.4 we show the DOS for chains made from various numbers of wells.

For small values of N (see Figure 2.3) the DOS shows N peaks distributed over the energy range. For long chains with $N = 1000$ (see Figure 2.4) the DOS is close to that of an infinite array: it changes on doubling the period from a single band to two sub-bands with a gap of width $|e_A - e_B|$ (top panels in Figure 2.4), while in the case of a Fibonacci array quasi-periodicity induces a fragmentation of the spectrum and the appearance of pseudo-gaps (bottom panel in Figure 2.4).

This is a typical result of disorder introduced by quasi-periodic or even aperiodic modulations of the site energies. In particular, in the case of a quasi-periodic Fibonacci chain the spectrum forms a Cantor set with zero measure. The emergence of minigaps in the DOS can cause particle localization as we shall illustrate in the Section 2.4.

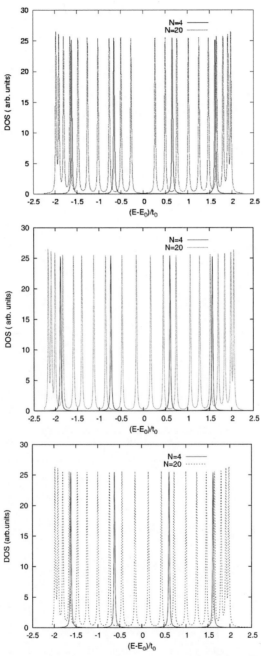

Figure 2.3. DOS of single-well array (top), double-well array (center), and Fibonacci-ordered array (bottom) for $N = 4$ (solid line) and $N = 20$ (dotted line), as a function of energy E referred to the center of the spectrum at energy E_0 and for spectral width $4t_0$.

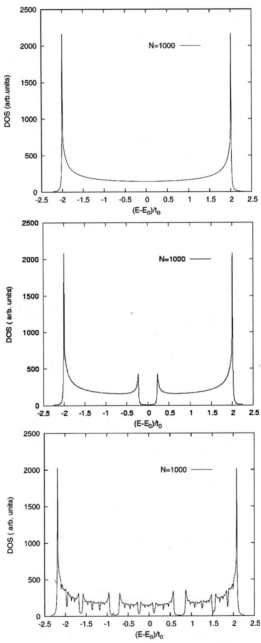

Figure 2.4. DOS as in Figure 2.3, for the case $N = 1000$.

2.3. Numerical results for transport by electrons through quantum-dot arrays

In this section, we take into account the presence of external leads to calculate DOS within Green's function scheme. We evaluate the dc transmission coefficients for an electron through 1D lattice and we show how the length of the array affects the transmission for different types of array.

2.3.1. The effects of the leads

In treating transport by electrons through an array of potential wells, we consider the arrays as connected to an incoming (l) and an outgoing (r) metallic lead. The leads are described by two additional terms in the Hamiltonian,

$$\mathcal{H}_{L,l} = \sum_{n=-\infty}^{0} \{ E_0^{in}|n\rangle\langle n| + t_0(|n\rangle\langle n+1| + \text{c.c.}) \} \qquad (2.9)$$

and

$$\mathcal{H}_{L,r} = \sum_{n=N+1}^{\infty} \{ E_0^{out}|n\rangle\langle n| + t_0(|n-1\rangle\langle n| + \text{c.c.}) \} . \qquad (2.10)$$

We consider below transport through the array by Fermi-surface electrons and, by imposing infinitesimal bias across the array. We also take the hopping energy to equal to one fourth of the spectral width.

The presence of the leads modifies the DOS and we accordingly have to decimate the leads and to consistently renormalize the energies of sites 1 and N in the chain. Equation (2.3) does not apply in this case and a modified Kirkman-Pendry relation as derived by Farchioni et al. [114] must be used when sites 1 and N are not edge sites. This relation is

$$D(E) = \frac{1}{\pi} \, \text{Im} \, \frac{\partial}{\partial\lambda} \ln G_{1,N}(E, \lambda) \Big|_{\lambda=0} , \qquad (2.11)$$

where the real parameter λ is introduced to select the sites on which the electronic states are counted. In our specific case $G_{1,N}(E, \lambda)$ can be written as

$$G_{1,N}(E, \lambda)$$

$$= \frac{\tilde{t}_{N,1}(\tilde{E})}{[\tilde{E} - \tilde{\varepsilon}_1^{(N-2)}(\tilde{E}) - \tilde{\mathcal{E}}(E)][\tilde{E} - \tilde{\varepsilon}_N^{(N-2)}(\tilde{E}) - \tilde{\mathcal{E}}(E)] - [\tilde{t}_{1,N}(\tilde{E})]^2} , \qquad (2.12)$$

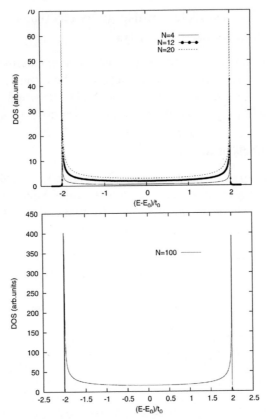

Figure 2.5. DOS of single-well chains connected to an incoming and an outgoing lead, as a function of energy E referred to the band center at energy E_0 and for a bandwidth of $4t_0$. Top panel: $N = 4$ (solid line), $N = 12$ (solid line and dots), and $N = 20$ (dashed line). Bottom panel: $N = 100$.

where $\tilde{E} = E + \lambda$ and the term $\tilde{\mathcal{E}}(E) = \frac{1}{2}(E - E_0) - [\frac{1}{4}(E - E_0)^2 - t_0^2]^{1/2}$ is the renormalized contribution of the leads to the energy of sites 1 and N [79].

The DOS of a single-well chain connected to leads is shown in Figure 2.5 for the case $E_0 = e_A$ and $t_0 = t_{A,A}$. This matching between chain and leads makes the DOS have the same features as for an infinite single-well array, independently of the number of sites. When matching is lost as in the double-well chain, the DOS strongly depends on the length of the chain as is shown in Figure 2.6. For a low number of wells ($N = 4$ or $N = 12$) the effect of the leads is dominant and the DOS resembles that of a 1D monatomic lattice. A depression is seen to emerge at the center of the spectrum at $N = 20$ and for a long array ($N = 100$) the effect of the leads becomes irrelevant and the DOS is very similar to that shown for the isolated array in the center panel of Figure 2.4.

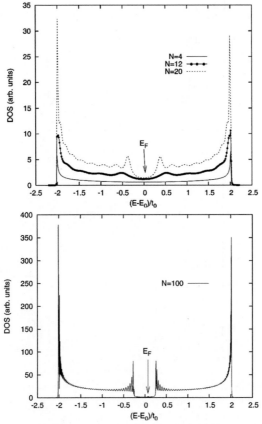

Figure 2.6. DOS as in Figure 2.5, for double-well chains connected to an incoming and an outgoing lead.

Similar effects are seen in Figure 2.7 on the DOS of the Fibonacci-ordered chains: at low N the structure of the DOS shown in Figure 2.3 is washed out by the presence of the leads, while at large N the typical features due to quasi-periodicity re-appear.

The location of the Fermi level becomes of course crucial for both the double-well and the Fibonacci-ordered chain. At half filling the Fermi energy E_F is located near E_0, as indicated in Figures 2.6 and 2.7. In this case the DOS at the Fermi level vanishes on increasing the length of the double-well chain, yielding a band insulator. In a Fibonacci-ordered array a disorder-induced insulating state will be met wherever the Fermi level is inside one of the pseudo-gaps shown in Figure 2.7 at large N.

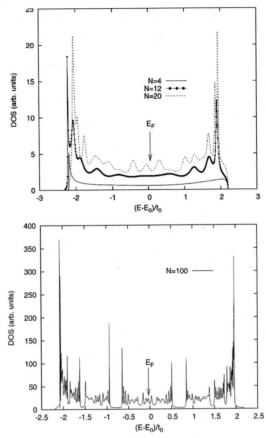

Figure 2.7. DOS as in Figure 2.5, for Fibonacci-ordered chains connected to an incoming and an outgoing lead.

2.3.2. The transmittivity

We rewrite the Hamiltonian $\mathcal{H} = \tilde{\mathcal{H}}_w + \mathcal{H}_{L,l} + \mathcal{H}_{L,r}$ of the effective dimer connected to the leads as the sum of two terms, $\mathcal{H} = \mathcal{H}_0 + \mathcal{H}_I$. The term $\mathcal{H}_0$ describes an infinite perfect chain with spacing d, site energy E_0, and hopping energy t_0, whose elements $(2, 3, \ldots, N-1)$ have been decimated as previously described. It is given by

$$\mathcal{H}_0 = \mathcal{H}_{L,l} + \mathcal{H}_{L,r} + \tilde{E}_0 \left(|1\rangle\langle 1| + |N\rangle\langle N|\right) + \tilde{t}_0 (|1\rangle\langle N| + |N\rangle\langle 1|). \quad (2.13)$$

The remainder is

$$\mathcal{H}_I = \tilde{\mathcal{H}}_w - \{\tilde{E}_0(|1\rangle\langle 1| + |N\rangle\langle N|) + \tilde{t}_0(|1\rangle\langle N| + |N\rangle\langle 1|)\} \quad (2.14)$$

and will be viewed as a perturbation determining scattering of incoming waves. In these equations the quantities $\tilde{E}_0$ and $\tilde{t}_0$ are obtained from equations (2.5) and (2.7) for $j = N - 2$ by taking $e_i = E_0$ and $t_{i,i+1} = t_0$ for all i.

The wavefunction $|\varphi\rangle$ at energy E in the continuous spectrum of $\mathcal{H}$ is obtained from the wavefunction $|k\rangle$ of the unperturbed periodic Hamiltonian $\mathcal{H}_0$, the unperturbed Green's function $G^0(E) = (E - \mathcal{H}_0)^{-1}$, and the T-matrix $T(E) = \mathcal{H}_1 (1 - G^0 \mathcal{H}_1)^{-1}$ as $|\varphi\rangle = |k\rangle + G^0 T |k\rangle$, T as well as $\mathcal{H}_1$ being 2×2 matrices in the space spanned by $|1\rangle$ and $|N\rangle$. The projection of $|\varphi\rangle$ onto the localized function $|n\rangle$ is

$$\langle n|\varphi\rangle = \langle n|k\rangle + \sum_{l,j=1,N} G^0_{n,l} \, T_{l,j} \, \langle j|k\rangle, \tag{2.15}$$

where $G^0_{n,l} = \langle n|G^0|l\rangle$, $T_{l,j} = \langle l|T|j\rangle$, and $\langle j|k\rangle = e^{ikjd}$. The expressions for these matrix elements can be found in [113]. We can then write equation (2.15) in the form

$$\langle n|\varphi\rangle = e^{iknd}$$
$$+\left(G^0_{N,1} T_{1,N} + G^0_{1,N} T_{N,1} e^{-2ik(N-1)d} + G^0_{N,N} T_{N,N} + G^0_{1,1} T_{1,1}\right) e^{iknd} \tag{2.16}$$

and define the transmittance τ and the reflectance ρ by setting

$$\langle n|\varphi\rangle = \begin{cases} \tau e^{iknd} & (n > N) \\ e^{iknd} + \rho \, e^{-iknd} & (n < 1) \end{cases}. \tag{2.17}$$

We thus obtain

$$\tau = 1 + G^0_{N,1} T_{1,N} + G^0_{1,N} T_{N,1} e^{-2ik(N-1)d} + G^0_{N,N} T_{N,N} + G^0_{1,1} T_{1,1}. \tag{2.18}$$

Notice the difference between this equation (2.18), which is valid for both real and effective dimers, and equation (8) in [113], which applies only to real dimers.

The transmission coefficient $\mathcal{T}$ is given by $\mathcal{T} = |\tau|^2$. In a single-well chain with $E_0 = e_A$ and $t_0 = t_{A,A}$, the perturbation Hamiltonian is null by construction and the transmission coefficient is equal to unity for any value of the energy inside the spectrum. Numerical results for the double-well array and the Fibonacci-ordered chain are shown in Figures 2.8 and 2.9.

The features of the transmittivity as a function of the energy parallel those of the DOS in Figures 2.6 and 2.7, where the effects of the leads have been included: the transmittivity drops as the DOS decreases and vanishes inside the energy gaps.

At half filling the transmittivity of the double-well array at the Fermi level is high for very short arrays but rapidly drops on increasing the number of wells, so that the array becomes insulating. On the contrary, the

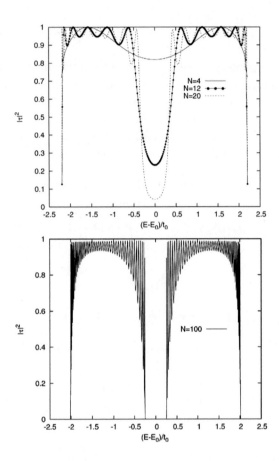

Figure 2.8. Transmission coefficient of a double-well chain, as a function of energy E referred to the band center at energy E_0 and for spectral width $4t_0$. Top panel: $N = 4$ (solid line), $N = 12$ (solid line and dots), and $N = 20$ (dashed line). Bottom panel: $N = 100$.

transmittivity near the Fermi level remains high in a Fibonacci-ordered array, so that metallic-like conduction is preserved independently of the length of the array. There are, however, strong minima in the transmittivity in correspondence to the pseudo-gaps in the DOS.

2.4. Numerical results for transport by atomic quantum gases through optical lattices

In this section, we describe how various types of array can be created from the interference of suitably arranged laser beams. Local density of states is introduced as a proper concept for the calculation of the transmission coefficient of a gas through an extended array. Finally, we we

understand the concept of an LDA we recall first how the noninteracting kinetic energy $T_s[n]$ is treated in the Thomas-Fermi approximation: In a *homogeneous* system one can write the kinetic energy per unit volume t_s as

$$t_s^{\text{hom}}(n) = \frac{3\hbar^2}{10m}(3\pi^2)^{2/3} n^{5/3}, \qquad (3.28)$$

where $n = \text{const}$. In an inhomogeneous system with $n = n(\mathbf{r})$ one approximates locally

$$t_s(\mathbf{r}) \approx t_s^{\text{hom}}(n)\Big|_{n \to n(\mathbf{r})} = \frac{3\hbar^2}{10m}(3\pi^2)^{2/3} n(\mathbf{r})^{5/3}, \qquad (3.29)$$

and obtains the full kinetic energy by integration over all space:

$$T_s^{\text{LDA}}[n(\mathbf{r})] = \int d\mathbf{r}\, t_s^{\text{hom}}(n(\mathbf{r})) = \frac{3\hbar^2}{10m}(3\pi^2)^{2/3}\int d\mathbf{r}\, n(\mathbf{r})^{5/3}. \quad (3.30)$$

For the kinetic energy the approximation $T_s[n] \approx T_s^{\text{LDA}}[n]$ is much inferior to the exact treatment of T_s in terms of orbitals, offered by the Kohn-Sham equations, but the LDA concept turned out to be highly useful for another component of the total energy (3.25), the exchange-correlation energy $E_{\text{xc}}[n]$.

With the same manner of replacing locally the constant density in the homogeneous case with the space-dependent density in the inhomogeneous case, we can employ the LDA for the E_{xc} functional by bulding it from the exchange-correlation energy $\varepsilon_{\text{xc}}(n)$ per particle of the homogeneous system taken at the local density, according to

$$E_{\text{xc}}[n] \approx E_{\text{xc}}^{\text{LDA}}[n(\mathbf{r})] = \int d\mathbf{r}\, n(\mathbf{r})\, \varepsilon_{\text{xc}}^{\text{hom}}(n(\mathbf{r})). \qquad (3.31)$$

This approximation for $E_{\text{xc}}[n]$ has proved successful, even when applied to systems that are quite different from the electron liquid that forms the *reference system* for the LDA. In the following of this thesis, we will use an LDA for the Hubbard model.

At this stage it may be worthwhile to recapitulate what practical DFT does, and where the LDA enters its conceptual structure: What real systems, such as actual molecules and solids, have in common, is that they are simultaneously inhomogeneous (the electrons are exposed to spatially varying electric fields produced by the nuclei) and interacting (the electrons interact *via* the Coulomb interaction). The way density-functional theory, in the local-density approximation, deals with this inhomogeneous many-body problem is by decomposing it into two simpler (but still highly nontrivial) problems: the solution of a spatially uniform many-body problem (the homogeneous electron liquid) yields the uniform exchange-correlation energy $\varepsilon_{xc}^{hom}(n)$, and the solution of a spatially inhomogeneous noninteracting problem (the inhomogeneous electron gas) yields the particle density. Both steps are connected by the local-density approximation, equation (3.31), which shows how the exchange-correlation energy of the uniform interacting system enters the equations for the inhomogeneous noninteracting system [127].

3.2. Spin-density-functional theory

As presented in Section 3.1, DFT was initially formulated for inhomogeneous electronic systems in their ground state. A number of extensions of the basic DFT approach have been developed over the years, covering electrons in magnetic fields, spin-polarized systems, Luttinger liquids, superconductors and superfluids, boson and fermion gases, and classical liquids. In this section we indicate the formulation of SDFT which is a generalization of density-functional theory by considering the spin-up and spin-down densities as separate variables. Its fundamental variables $n_\uparrow(\mathbf{r})$ and $n_\downarrow(\mathbf{r})$ can be used to calculate the charge density $n(\mathbf{r})$ and the spin-magnetization $m(\mathbf{r})$ from

$$n(\mathbf{r}) = n_\uparrow(\mathbf{r}) + n_\downarrow(\mathbf{r}), \qquad (3.32)$$

$$m(\mathbf{r}) = n_\uparrow(\mathbf{r}) - n_\downarrow(\mathbf{r}). \qquad (3.33)$$

More generally, the Hohenberg-Kohn theorem of SDFT states that the ground-state wave function and all ground-state observables are unique functionals of n and m or equivalently of $n_\uparrow$ and $n_\downarrow$ [129].

In SDFT the coupling with the external field $B(\mathbf{r})$ is given by a Zeeman term that is different for spin-up and spin-down in the last term in Hamiltonian (3.1)

$$\hat{\mathcal{H}}_1 = \sum_\sigma \int d\mathbf{r} \, W_\sigma(\mathbf{r}) \hat{\psi}_\sigma^\dagger(\mathbf{r}) \hat{\psi}_\sigma(\mathbf{r}), \qquad (3.34)$$

where

$$W_\sigma(\mathbf{r}) = V_{ext}(\mathbf{r}) - \gamma \, \text{sgn}(\sigma) \, B(\mathbf{r}). \qquad (3.35)$$

Here sgn is the sign function, $\hat{\psi}_\sigma^\dagger(\mathbf{r})\hat{\psi}_\sigma(\mathbf{r})$ is the spin-density operator $\hat{n}_\sigma(\mathbf{r})$ and $\gamma = g\mu_B/2$ (with g the Landé factor and μ_B the Bohr magneton). Of course, the scalar external potential and the external magnetic field are now arbitrarily strong. Provided that the spin densities may be varied independently, the formal aspects of the theory are a straightforward extension of standard DFT: one only needs to introduce energy functionals that depend on a multiplicity of variables (the spin densities) and proceed to derive self-consistent Kohn-Sham equations for spin-dependent single-particle orbitals. The properties of the E_{xc} functional in SDFT have been critically examined by Capelle and Vignale [130]: a non-uniqueness problem arises at those densities where the ground state is rigid against combinations of the external fields, and at these densities the E_{xc} functional presents derivative discontinuities.

It is nevertheless true that SDFT has in many recent studies replaced the standard DFT approach even in the treatment of non-magnetic systems. In essence, the increased number of density variables allows more flexibility in the search of the energy minimum and an accurate determination of the local spin densities.

3.3. Lattice density-functional theory

Despite the success of DFT, some fundamental problems remain, like the question of the continuity of the exchange-correlation potential [131, 132]. In order to partly simplify this issue, the equivalent of DFT on a lattice has been introduced, where the local site occupancies are treated as the basic variables. This section deals with the main problem of continuity and lattice density-functional theory.

3.3.1. The "band-gap" problem in the physics of semiconductors and insulators

One of the most intriguing properties of the exact functional, which has resisted all attempts of describing it in local approximations, is the derivative discontinuity of the exchange-correlation functional with respect to the total particle number. This discontinuity of the exchange-correlation potential is defined as [126]

$$\Delta_{xc} \equiv V_{xc}^{(+)} - V_{xc}^{(-)} = \left. \frac{\delta E_{xc}[n]}{\delta n(\mathbf{r})} \right|_{N+\delta} - \left. \frac{\delta E_{xc}[n]}{\delta n(\mathbf{r})} \right|_{N-\delta}, \qquad (3.36)$$

where δ is an infinitesimal shift of the electron number N, and Δ_{xc} is a system-dependent, but $\mathbf{r}$-independent shift of the exchange-correlation potential Δ_{xc} as it passes from the electron-poor to the electron-rich side of integer N. This quantity is of partial importance for the discussion of the band-gap in an insulator or semiconductor. The band-gap is rigorously defined as the difference between the lowest conduction-band energy and the highest valance-band energy. The latter is the energy required to remove an electron from the insulating N-particle ground state to infinity; the former is obtained by adding an electron to the insulating

N-particle ground state. Thus in terms of ionisation potential

$$I = E(N - 1) - E(N),\qquad(3.37)$$

and the electron affinity

$$A = E(N) - E(N + 1),\qquad(3.38)$$

the true fundamental gap Δ is given by

$$\Delta = I - A = E(N + 1) + E(N - 1) - 2E(N).\qquad(3.39)$$

As an example we consider a noninteracting system with insulating ground state. The difference $I - A$ is then readily calculated as

$$\Delta_{\text{nonint}} = \varepsilon_{N+1}(N) - \varepsilon_N(N),\qquad(3.40)$$

where $\varepsilon_m(M)$ denotes the m-th single-particle level of the M-particle system. Returning to the interacting case, both I and A can be expressed in terms of the chemical potential $\mu(N)$ to give [126]

$$\Delta = -\mu(N - \delta) + \mu(N + \delta).\qquad(3.41)$$

It can be shown that the band gap is rigorously given by

$$\Delta = \left\{ \frac{\delta T_s[n]}{\delta n(\mathbf{r})}\bigg|_{N+\delta} - \frac{\delta T_s[n]}{\delta n(\mathbf{r})}\bigg|_{N-\delta} \right\}_{n=n_0}$$
$$+ \left\{ \frac{\delta E_{\text{xc}}[n]}{\delta n(\mathbf{r})}\bigg|_{N+\delta} - \frac{\delta E_{\text{xc}}[n]}{\delta n(\mathbf{r})}\bigg|_{N-\delta} \right\}_{n=n_0}.\qquad(3.42)$$

This fundamental derivative has to be evaluated at the ground state density $n(\mathbf{r})$ of the N-particle insulator.

For noninteracting systems, equation (3.42) reduces to

$$\Delta_{\text{nonint}} = \frac{\delta T_s[n]}{\delta n(\mathbf{r})}\bigg|_{N+\delta} - \frac{\delta T_s[n]}{\delta n(\mathbf{r})}\bigg|_{N-\delta}.\qquad(3.43)$$

If evaluated at the interacting ground state density $n(\mathbf{r})$, the right-hand side of equation (3.43) yields the Kohn-Sham gap

$$\Delta_{\text{KS}} = \left\{ \frac{\delta T_s[n]}{\delta n(\mathbf{r})}\bigg|_{N+\delta} - \frac{\delta T_s[n]}{\delta n(\mathbf{r})}\bigg|_{N-\delta} \right\}_{n=n_0},\qquad(3.44)$$

which by (3.40) is identical with

$$\Delta_{KS} = \varepsilon_{N+1}^{KS}(N) - \varepsilon_N^{KS}(N). \tag{3.45}$$

Combining equation (3.42) with equations (3.36) and (3.44), one finally obtains the exact representation

$$\Delta = \Delta_{KS} + \Delta_{xc}. \tag{3.46}$$

If, for an insulator, the Kohn-Sham equations are solved using the local-density approximation for V_{xc}, the resulting Kohn-Sham gap Δ_{KS}^{LDA} is found to be too small in comparison with experiment (see Table 3.1).

Table 3.1. Band gaps (eV) of selected semiconductors and insulators: LDA vs experiments. Adapted from [41].

	Diamond	Si	Ge	LiCl	GaAs
LDA	3.9	0.52	0.07	6.0	0.12
Expt.	5.48	1.17	0.744	9.4	1.52

Since equation (3.46) is an exact representation of the experimental gap, the error found for Δ_{KS}^{LDA} has two distinct sources: (i) the discontinuity Δ_{xc} is neglected; (ii) the value Δ_{KS}^{LDA} obtained with the local density approximation is not equal to the exact Kohn-Sham gap Δ_{KS}. Recent work has therefore been focused on the question whether (i) or (ii) is the dominant effect.

This question is of tremendous practical importance: If the neglected discontinuity is the main source of error, then any attempt to improve the gap by going beyond the local density approximation is bound to fail. A partial answer was given by Godby, Schlüter, and Sham [133] who calculated a potential $V_{xc}(\mathbf{r})$ which can be expected to agree very closely with the true exchange-correlation potential. This potential turned out to be in remarkably close agreement with the local density approximation. It therefore appears likely that Δ_{KS}^{LDA} agrees very closely with the true Kohn-Sham gap Δ_{KS}. As a consequence, the neglected discontinuity should be the principal source of error. (In silicon, for example, Δ_{xc} is responsible for over 80% of the LDA-gap error) [126].

However, Gunnarsson and Schönhammer [134, 135] found a very small discontinuity in a simple, one dimensional, Hubbard-like semiconductor model. Here the exchange-correlation potential and its discontinuity can be calculated exactly (see below). In the following section we introduce the homogeneous Hubbard model that later will be used as a reference system to deal with the inhomogeneous system.

3.3.2. The homogeneous Hubbard model as a reference system

The Hubbard model is named after John Hubbard, who in a series of influential articles [136, 137] introduced a Hamiltonian in order to model electronic correlations and proposed a number of approximate treatments of the associated many-body problem. There is a wide range of properties in condensed matter systems that were or are under investigation by modeling them by the Hubbard model: ferromagnetism, conductor-insulator transitions and high-T_c superconductivity [138].

In the original version, Hubbard model is a model of itinerant, interacting electrons on a lattice. The structure and the dimension of the lattice influence its features. An *exact* solution of the model is only known in 1D [122] (see Section 3.4). In the rest of this thesis we focus on the 1D Hubbard model.

The defining features of the simplest formulation of the Hubbard model are:

(i) The electrons are strongly localized at the sites of the lattice. This means that the electron field operator is given by $\psi(z_i)$, where z_i label the lattice sites, instead of some continuous operator $\psi(z)$.

(ii) The Coulomb interaction between electrons at different lattice sites is neglected. Any electron interacts only with a possible second electron at the same lattice site. Due to the Pauli principle, only two electrons with opposite spin at one lattice site are allowed.

(iii) The electrons have the ability to hop between lattice sites. Electron hopping occurs only between nearest neighbor lattice sites. Furthermore, we assume that the hopping amplitude is the same for all nearest neighbor pairs.

With these assumptions, we can write down the Hamiltonian for the single-band Hubbard model as

$$\hat{\mathcal{H}} = \hat{\mathcal{H}}_{\text{hop}} + \hat{\mathcal{H}}_{\text{int}}$$
$$= \sum_{\langle i,j \rangle} \sum_{\sigma=\uparrow,\downarrow} t_{ij} \left[\hat{c}_\sigma^\dagger(z_i)\hat{c}_\sigma(z_j) + \text{H.c.} \right] + U \sum_i \hat{n}_\uparrow(z_i)\hat{n}_\downarrow(z_i), \quad (3.47)$$

where $\hat{c}_\sigma^\dagger(z_i)$ and $\hat{c}_\sigma(z_i)$ are, respectively, the creation and annihilation operators for a fermion atom of spin-1/2 degree of freedom $\sigma = \uparrow, \downarrow$ in the Wannier state at the i-th lattice site and $\hat{n}_\sigma(z_i) = \hat{c}_\sigma^\dagger(z_i)\hat{c}_\sigma(z_i)$ is the spin-resolved site occupation operator normalized to the number of particles with spin σ, $N_\sigma = \langle \sum_i \hat{n}_\sigma(z_i) \rangle$, and $\hat{n}(z_i) = \sum_{i,\sigma} \hat{n}_\sigma(z_i) = \hat{n}_\uparrow(z_i) + \hat{n}_\downarrow(z_i)$ is the total site occupation operator with

$$N = \left\langle \sum_{i,\sigma} \hat{n}_\sigma(z_i) \right\rangle = N_\uparrow + N_\downarrow, \tag{3.48}$$

and the local spin also is defined as

$$\hat{s}(z_i) = \frac{\hat{n}_\uparrow(z_i) - \hat{n}_\downarrow(z_i)}{2}. \tag{3.49}$$

The first term in equation (3.47) describes the hopping between different lattice sites. With our assumptions $t_{ij} = -t$ if i and j are nearest-neighbor sites and zero otherwise. Therefore the summation (indicated by $\langle i, j \rangle$) is over nearest neighbors, and one often considers periodic boundary condition, which means that $\langle i, j \rangle$ includes a term coupling opposite edges of the lattice. The sign of t is purely conventional. It can be shown [122, 139] that the energy spectrum of equation (3.47) is invariant under the replacement t by $-t$. The second term of the Hamiltonian describes the local Coulomb interaction between electrons at the same lattice site. In this Chapter, we take $U > 0$ to have a repulsive interaction ($U > 0$ raises the energy of placing two electrons on the same lattice site, which corresponds to a repulsive force). In Chapter 4 we will deal with $U < 0$ which corresponds to an attractive interaction.

If we consider also an external static potential applied to the moving fermions on a periodic lattice with length L, we can describe the system by a single-band inhomogeneous Hubbard model:

$$\begin{aligned}
\hat{\mathcal{H}} = &-t \sum_{i,\sigma} \left[\hat{c}_\sigma^\dagger(z_i)\, \hat{c}_\sigma(z_{i+1}) + \text{H.c.} \right] \\
&+ U \sum_i \hat{n}_\uparrow(z_i)\hat{n}_\downarrow(z_i) + \sum_{i,\sigma} V_{\text{ext}}^\sigma(z_i)\, \hat{n}_\sigma(z_i),
\end{aligned} \tag{3.50}$$

with $V_{\text{ext}}^\sigma(z_i)$ denoting the external confining potential. In this thesis we consider only a purely harmonic confining potential of the general form

$$V_{\text{ext}}^\sigma(z_i) = V_2^\sigma\, (z_i - L/2)^2 , \tag{3.51}$$

where V_2^σ is the strength of the trap for atoms with spin σ.

The Hamiltonian (3.50) will be the object of our theoretical investigation in the following sections. In the absence of external potential, the Hubbard Hamiltonian reduces to the homogeneous model which has exact solution as provided by Lieb and Wu [122]. These results will be explained in the Section 3.4 and we will use them to construct the exchange-correlation energy of the homogeneous case. This enables us to treat the trapped gas within a local-spin-density approximation.

3.3.3. Site-occupation functional theory

The aim of this section is to present a summary of the two key results of SOFT [134, 135, 140, 141]: (i) the Hohenberg-Kohn theorem and (ii) the Kohn-Sham mapping to an auxiliary noninteracting system.

The basic variable of SOFT is the site-occupation $n(z_i) = \langle \Psi | \hat{n}(z_i) | \Psi \rangle$, where $|\Psi\rangle$ is a generic many-body state. As in standard DFT, the central result of SOFT is the Hohenberg-Kohn theorem, which can be summarized in three key statements: (a) the ground state expectation value of any observable $\hat{\mathcal{O}}$ is a unique functional $\mathcal{O} = \langle GS | \hat{\mathcal{O}} | GS \rangle = \mathcal{O}[n]$ of the ground state site-occupation $n(z_i)$; (b) the ground state site-occupation minimizes the total-energy functional $E_{GS}[n]$; and (c) $E_{GS}[n]$ can be written as

$$E_{GS}[n] = F_{HK}[n] + \sum_i V_{ext}(z_i)\, n(z_i)\,, \qquad (3.52)$$

where $F_{HK}[n] = \langle \Psi | \hat{\mathcal{H}}_{hop} + \hat{\mathcal{H}}_{int} | \Psi \rangle$ is a *universal* functional of the site occupation, in the sense that it does not depend on the external potential.

Part (b) of the HK theorem suggests that if the exact analytical expression of $F_{HK}[n]$ is known, the ground state energy and the ground state site occupation could be found by solving the Euler-Lagrange equation

$$\frac{\delta F_{HK}[n]}{\delta n(z_i)} + V_{ext}(z_i) = \text{const}\,, \qquad (3.53)$$

the constant having the meaning of a Lagrange multiplier to enforce particle-number conservation.

The Kohn-Sham mapping, again in analogy with standard DFT, provides an essential simplification. One considers a noninteracting auxiliary system described by the Hamiltonian

$$\hat{\mathcal{H}}_s = -\sum_{i,j}\sum_{\sigma} t_{ij} \left[\hat{c}_\sigma^\dagger(z_i)\hat{c}_\sigma(z_j) + \text{H.c.} \right] + \sum_i V_{KS}(z_i)\, \hat{n}(z_i)\,. \quad (3.54)$$

The central assertion used in establishing the mapping is that for any interacting system there exists a local single-particle potential $V_{KS}(z_i)$ such that the exact ground-state site occupation $n(z_i)$ of the interacting system equals the ground-state site occupation of the auxiliary problem $n(z_i) = n^{(s)}(z_i)$ (noninteracting v-representability). According to part (c) of the Hohenberg-Kohn theorem there then exists a unique energy functional $E_s[n] = T_s[n] + \sum_i V_{KS}(z_i) n(z_i)$, for which the variational equation $\delta E_s[n] = 0$ yields the exact ground-state site occupation $n^{(s)}(z_i)$ corresponding to $\hat{\mathcal{H}}_s$. $T_s[n]$ denotes the universal kinetic energy functional of noninteracting pseudospin-$\frac{1}{2}$ fermions.

Suppose that the ground state of $\hat{\mathcal{H}}_s$ is nondegenerate. The ground state site occupation $n^{(s)}(z_i)$ (and thus, by assumption, $n(z_i)$) possesses a unique representation

$$n(z_i) = \sum_{\alpha, \text{occ.}} |\varphi_\alpha(z_i)|^2 , \qquad (3.55)$$

in terms of the lowest N single-particle orbitals obtained from the Kohn-Sham-Schrödinger equation

$$\sum_j \left[-t_{ij} + V_{KS}(z_i)\, \delta_{ij} \right] \varphi_\alpha(z_j) = \varepsilon_\alpha \varphi_\alpha(z_i) . \qquad (3.56)$$

Once the existence of a potential $V_{KS}(z_i)$ generating $n(z_i)$ via equations (3.55) and (3.56) is assumed, uniqueness of $V_{KS}(z_i)$ follows from the Hohenberg-Kohn theorem. Thus the single-particle orbitals $\varphi_\alpha(i) = \varphi_\alpha[n](z_i)$ are unique functionals of $n(z_i)$, and the noninteracting kinetic energy

$$T_s[n] = -\sum_{i,j} \sum_\alpha \left(\varphi_\alpha^*(z_i) \varphi_\alpha(z_j) + \text{c.c.} \right) , \qquad (3.57)$$

is a unique functional of $n(z_i)$ as well.

It is convenient at this point to write the total energy functional $E_{GS}[n]$ in equation (3.52) by introducing $T_s[n]$ and a Hartree term $E_H = U \sum_i n^2(z_i)/4$, i.e.

$$E_{GS}[n] = T_s[n] + E_H[n] + E_{xc}[n] + \sum_i V_{ext}(z_i)\, n(z_i) , \qquad (3.58)$$

or practically similar to the equation (3.27) as

$$E_{GS}[n] = \sum_\alpha \varepsilon_\alpha - \sum_i V_{xc}(z_i)\, n(z_i) - \sum_i V_H(z_i) + E_{xc} . \qquad (3.59)$$

where the exchange-correlation functional is formally defined as $E_{xc}[n] \equiv F_{HK}[n] - T_s[n] - E_H[n]$. The HK variational principle ensures that $E[n]$ is stationary under small variations $\delta n(z_i)$ around $n(z_i)$:

$$E_{GS}[n + \delta n] - E_{GS}[n]$$
$$= \delta T_s + \sum_i \delta n(z_i) \left[V_{ext}(z_i) + \frac{U}{2} n(z_i) + V_{xc}[n](z_i) \right] = 0 , \qquad (3.60)$$

where $V_{xc}[n](z_i)$ denotes the exchange-correlation potential,

$$V_{xc}[n](z_i) = \frac{\delta E_{xc}[n]}{\delta n(z_i)}\bigg|_{n(z_i)}. \tag{3.61}$$

Using $\delta T_s = -\sum_i V_{KS}(z_i)\,\delta n(z_i)$, we find that the Kohn-Sham potential is given by

$$V_{KS}(z_i) = V_{H}[n](z_i) + V_{xc}[n](z_i) + V_{ext}(z_i), \tag{3.62}$$

where $V_{H}[n](z_i) = U\,n(z_i)/2$.

3.4. The Bethe-Ansatz solution of the 1D homogeneous Hubbard model

In 1968 Lieb and Wu discovered [122] that the Bethe-Ansatz can be applied to the Hubbard model. They reduced the spectral problem of the Hamiltonian to solving a set of algebraic equations, known as the Lieb-Wu (LW) equations. In this section we mention these equation which gives the ground-state energy of the homogeneous Hubbard model.

For the 1D Hubbard model equation (3.50) in the absence of external potential, the commutation relations

$$\left[\sum_i \hat{n}_{\uparrow}(z_i), \hat{\mathcal{H}}\right] = \left[\sum_i \hat{n}_{\downarrow}(z_i), \hat{\mathcal{H}}\right] = 0, \tag{3.63}$$

imply that the numbers of spin-down fermions $N_{\downarrow}$ and spin-up fermions $N_{\uparrow}$ are good quantum numbers. Therefore we characterize the eigenstates by $N_{\downarrow}$ and $N_{\uparrow}$, and write the Schrödinger equation as

$$\hat{\mathcal{H}}\,|N_{\downarrow}, N_{\uparrow}\rangle = E(N_{\downarrow}, N_{\uparrow})\,|N_{\downarrow}, N_{\uparrow}\rangle. \tag{3.64}$$

In the thermodynamic limit where the lattice length L and the total number of fermions N become large ($L, N \to \infty$), the properties of the 1D Hubbard model are determined by the band filling n which is the number of fermions per lattice site, the spin s and the dimensionless coupling u. They are defined as

$$n = \frac{N_{\uparrow} + N_{\downarrow}}{L} = \frac{N}{L}, \tag{3.65}$$

$$s = \frac{N_{\uparrow} - N_{\downarrow}}{2L}, \tag{3.66}$$

$$u = \frac{U}{t}, \tag{3.67}$$

in which $0 \le n \le 1$ and $0 \le s \le \frac{n}{2}$. The limiting cases $n = 0$ and $n = 1$ correspond to empty and half-filled bands respectively.

According to Lieb and Wu [122], the ground state of the repulsive 1D homogeneous Hubbard model in the thermodynamic limit is described by two continuous distribution functions $\rho(x)$ and $\sigma(y)$ which satisfy the Bethe-Ansatz (BA) coupled Fredholm integral equations,

$$\rho(x) = \frac{1}{2\pi} + \frac{\cos x}{\pi} \int_{-B}^{+B} \frac{u/4}{(u/4)^2 + (y - \sin x)^2} \sigma(y) \, dy, \quad (3.68)$$

$$\sigma(y) = \frac{1}{\pi} \int_{-Q}^{+Q} \frac{u/4}{(u/4)^2 + (y - \sin x)^2} \rho(x) \, dx$$

$$- \frac{1}{\pi} \int_{-B}^{+B} \frac{u/2}{(u/2)^2 + (y - y')^2} \sigma(y') \, dy'. \quad (3.69)$$

The parameter Q is determined by the normalization condition

$$\int_{-Q}^{Q} \rho(x) \, dx = n, \quad (3.70)$$

while $\sigma(y)$ is normalized according to

$$\int_{-B}^{B} \sigma(y) \, dy = \frac{n}{2} - s. \quad (3.71)$$

The ground state energy of the system (per site) is given by

$$\varepsilon_{GS}(n \leq 1, s, u) = -2t \int_{-Q}^{Q} dx \, \rho(x) \cos x. \quad (3.72)$$

From right-hand-side of equation (3.71) an important constraint enters the solution which is

$$n > 2s. \quad (3.73)$$

This condition can be understood from the equations (3.65) and (3.66): both filling n and spin s are linear combinations of spin-up and spin-down number of fermionic atoms, so when we fix one of them we can not arbitrarily vary the other one.

For the repulsive side the study made by Schulz [142] shows that 1D Hubbard model describes a Luttinger liquid if $n \neq 1$ or 2. At half-filling i.e. $n = 1$, the ground state is a Mott insulator for every $u \neq 0$ (see below), while for $n = 2$ it is a band insulator. Two metallic ground state branches for $n > 1$ and $n < 1$ are connected by particle-hole symmetry,

$$\varepsilon_{GS}(n > 1, s; u) = (n - 1)u + \varepsilon_{GS}(2 - n, s; u), \quad (3.74)$$

Notice that the presence of a Mott insulating ground state at half-filling is signaled by a cusp in the ground state at $n = 1$, induced by the linear term in this relation (see Figure 3.2). Correspondingly the charge excitations spectrum possesses a gap.

The equations (3.68)-(3.72) are the main results by LW solution for the repulsive 1D Hubbard model. This set of equations plus the particle-hole symmetry relation equation (3.74) consist our main tool to obtain the results for ground state energy in the homogeneous case. In the case $s = 0$ we have $Q = \pi$ for all values of u, t and n. Figures 3.1 and 3.2 show the ground-state energy of the homogeneous Hubbard model for $s = 0$ and $s \neq 0$ respectively.

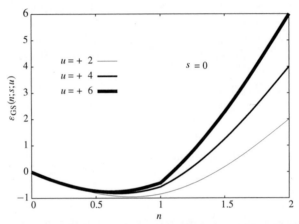

Figure 3.1. Ground state energy of repulsive 1D homogeneous Hubbard model (in units of hopping parameter t) as a function of the filling factor n for various values of the coupling parameter u. The cusp at $n = 1$ signals the Mott-insulating phase.

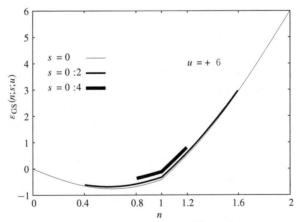

Figure 3.2. Ground state energy of repulsive 1D homogeneous Hubbard model (in units of hopping parameter t) as a function of the filling factor n for various spin s. The cusp at $n = 1$ signals the Mott-insulating phase.

Solution of the half-filled band

In the case of half-filled band we have $N = L$. This condition implies $Q = \pi$ and $B = \infty$. In this case the exact form for distribution functions $\rho(x)$ and $\sigma(y)$ can be derived [139]. We use the notation $\rho_0(x)$ and $\sigma_0(y)$ for these solutions. They can be read as

$$\sigma_0(y) = \frac{1}{2\pi} \int_0^\infty \frac{J_0(x)\,\cos(x\,y)}{\cosh(x\,u/4)}\,dx\,, \tag{3.75}$$

and

$$\rho_0(x) = \frac{1}{2\pi} + \frac{\cos x}{\pi} \int_0^\infty \frac{J_0(y)\,\cos(y\,\sin x)}{1 + \exp(y\,u/2)}\,dy\,. \tag{3.76}$$

The substitution of these distribution functions finally yields the ground state energy of the half-filled band as

$$\varepsilon_{GS}\,(n = 1, u) = -4t \int_0^\infty \frac{J_0(x)\,J_1(x)}{x\left[1 + \exp(x\,u/2)\right]}\,dx\,, \tag{3.77}$$

where $J_0(x)$ and $J_1(x)$ are, respectively, the Bessel functions of order zero and one.

3.4.1. Luttinger liquid behavior and the metal-insulator transition

In general there are two control parameters in the Hubbard model that can be varied: the interaction and the filling. Correspondingly two different Mott transitions exist: one can either stay at a given (commensurate) filling and vary the strength of the interactions (this transition is called the Mott-U), one can also keep the strength of the interaction constant and dope the system to move away from the commensurate density (a Mott-δ transition, a situation realized e.g in high T_c superconductors) [143]. In this section we will show, following Schulz [142] that the 1D Hubbard model away from half filling and full filling is in the Luttinger liquid universality class. We will then illustrate following Lieb and Wu how the 1D Hubbard model does not exhibit the Mott-U transition.

Luttinger liquid behavior

The Lieb-Wu equations (3.68)-(3.72) provide a complete picture of the energy spectrum of the 1D Hubbard model. On the other hand, the relevance of correlation effects in interacting fermion systems to high-T_c superconductivity [138] has led to growing interest in the computation of correlation functions for the 1D Hubbard model, since it combines the essentials of correlated electrons with the attractive feature of complete integrability. Unfortunately the wavefunctions obtained by LW are prohibitively complicated [144], and no practical scheme for the evaluation of expectation values or correlation functions has been found [145].

A relation between the spectrum and correlation exponents was found by Haldane [146]. He showed that this relation is a general property of a universal class of Luttinger liquids (see Section 1.1.4) which covers the soluble models *via* Bethe-Ansatz. As an example he argued and showed the $s = \frac{1}{2}$ Heisenberg-Ising-XY spin chain (or XXZ model) in a field belongs to this universality class and calculated the phase diagram and the correlation exponents [146].

By a straightforward generalizing of Haldane's idea, Schulz demonstrated that the Hubbard model realizes a Luttinger liquid. He made a connection [142] between the Hubbard model and the Luttinger liquid Hamiltonian in equation (1.78),

$$ H = H_\rho + H_\sigma + \frac{2g_1}{(2\pi A)^2} \int dx \cos\left(\sqrt{8}\phi_\sigma(x)\right), \qquad (3.78) $$

where $H_i = \rho, \sigma$ is given in equation (1.79). The so-called Bosonization scheme gives the Luttinger liquid parameter for the Hubbard model which are all functions of single variable U with [44]

$$ u_\rho K_\rho = u_\sigma K_\sigma = v_F, $$

$$ u_\rho/K_\rho = v_F\left(1 + \frac{U}{\pi v_F}\right), $$

$$ u_\sigma/K_\sigma = v_F\left(1 - \frac{U}{\pi v_F}\right), $$

$$ g_1 = U. \qquad (3.79) $$

For the repulsive side of the Hubbard model ($K_\rho < 1$) the original study made by Schulz extracted the velocities u_ρ and u_σ directly from the Bethe-ansatz, and obtained K_ρ from the compressibility. In particular K_ρ determines the long-distance decay of all the correlation functions of the system. For example the charge correlation function is [142]

$$ \langle n(x)\, n(0) \rangle = \frac{K_\rho}{(\pi x)^2} + A_1 \cos(2k_F x)\, x^{-1-k_\rho}\, \ln^{-3/2}(x) $$
$$ + A_2 \cos(4k_F)\, x^{-4K_\rho}, \qquad (3.80) $$

with the model-dependent constants A_1, A_2.

The values of the Luttinger liquid parameters are shown in Figures 3.3 and 3.4. Results for velocities $u_{\rho,\sigma}$ are shown for various values of U/t.

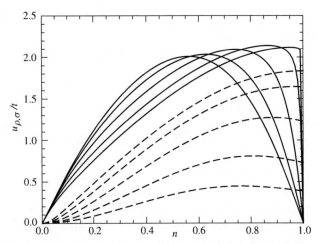

Figure 3.3. The charge and spin velocities u_ρ (full line) and u_σ (dashed line) for the Hubbard model, as a function of the band filling for different values of U/t: for u_σ $U/t = 1, 2, 4, 8, 16$ from top to bottom, for u_ρ $U/t = 16, 8, 4, 2, 1$ from top to bottom in the left part of the figure. From Schulz [145].

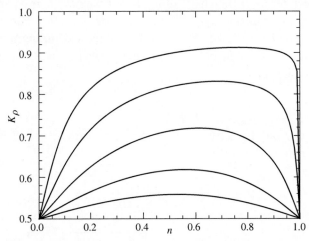

Figure 3.4. The correlation exponent K_ρ as a function of the band filling for different values of U: ($U/t = 1, 2, 4, 8, 16$ for the top to bottom curves). Note the rapid variations near $n = 1$ for small U. From Schulz [145].

At very small filling, the charge velocity goes linearly to zero and the spin velocity goes to a constant given by $J = 4t^2/U$ [44]. Note that for $U = 0$ one has $u_\rho = u_\sigma = 2t\,\sin(\pi\frac{n}{2})$, whereas for $U \to \infty$

$$u_\rho = 2t\,\sin(\pi n), \qquad (3.81)$$

$$u_\sigma = 2\pi\,\frac{t^2}{U}\left[\frac{1 - \sin(2\pi n)}{2\pi n}\right]. \qquad (3.82)$$

In the non-interacting case $u_\sigma \propto n$ for small n, but for *any* positive U $u_\sigma \propto n^2$ [145].

The parameter K_ρ also is obtained from the exact solution [142]. The results for K_ρ as a function of particle density are shown in Figure 3.4 for different values of U/t. The repulsive Hubbard model can only have a Mott phase at half-filling. At half-filling the system is an insulator for any repulsive interaction. At very small filling, $K_\rho \to \frac{1}{2}$ in agreement with the universal value of the Mott-δ transition [44].

Absence of a Mott transition

In the Hubbard model one inquires whether a Mott transition occurs at some critical filling. The strategy is to compute the chemical potential μ_+ (μ_-) for adding (removing) one electron at half filling. *The system is conducting if $\mu_+ = \mu_-$ and insulating if $\mu_+ > \mu_-$.*

For less than half-filling, $n < 1$, Lieb and Wu calculated μ_+ and μ_- and found that $\mu_+ = \mu_-$ for *all u i.e.* the chemical potential cannot make any jump in this region and the system is *always* in metallic phase. On the other hand, for more than half-filling, $n > 1$ we have to use particle-hole symmetry to calculate $\varepsilon_{GS}(n)$. The derivatives of $\varepsilon_{GS}(n)$, namely μ_+ and μ_-, can now be different above and below the half-filling point $n = 1$ and this gives rise to the possibility of having an insulator. We learn from equation (3.74) that

$$\mu_+ + \mu_- = u, \tag{3.83}$$

and hence $\mu_+ > \mu_-$ if $\mu_- < u/2$.
The calculation of μ_- can be done in closed form for a half-filled band with the result [122, 139]

$$\mu_-(u) - 2 = -4t \int_0^\infty \frac{J_1(x)}{x\left[1 + \exp(x\,u/2)\right]}\,dx$$

$$= -4t \sum_{n=1}^\infty (-1)^n \left[\sqrt{1 + \frac{n^2 u^2}{4}} - \frac{n u}{2}\right]. \tag{3.84}$$

It can be established from equations (3.83) and (3.84) that, indeed, $\mu_+(u) > \mu_-(u)$ for $u > 0$, and

$$\lim_{u \to 0} \mu_\pm = 0. \tag{3.85}$$

Therefore the 1D Hubbard model is insulating for all $u > 0$. In other word, the ground state for a half-filled band is insulating for any non-zero u, and conducting for $u = 0$.

3.5. Bethe-Ansatz-based local-spin-density approximation for the exchange-correlation potential

In this section we consider a spin-resolved confined repulsively inter-acting Fermi gas in 1D optical lattice. To obtain the ground-state density profiles of this inhomogeneous (trapped) model through SOFT, we employ the local-spin-density approximation (LSDA) for the exchange-correlation potential. As the *reference system* for LSDA we use the exact Lieb-Wu solution of the homogeneous Hubbard model which we presented in the previous section.

Generalizing the main SOFT scheme, presented in Section 3.3.3, to the spin-resolved Fermi gas, the exact ground-state site occupation for every species

$$n_\sigma(z_i) = \langle GS | \hat{n}_\sigma(z_i) | GS \rangle, \tag{3.86}$$

can be obtained by solving self-consistently the lattice Kohn-Sham equations for both spin-up and spin-down species separately

$$\sum_j \left[-t_{ij} + V_{ks}^\sigma[n_\uparrow, n_\downarrow](z_i)\, \delta_{ij} \right] \varphi_\sigma(z_j) = \varepsilon_\sigma \, \varphi_\sigma(z_i), \tag{3.87}$$

with Kohn-Sham effective potential

$$V_{KS}^\sigma[n_\uparrow, n_\downarrow](z_i) = V_{ext}^\sigma(z_i) + V_H^\sigma(z_i) + V_{xc}^\sigma(z_i), \tag{3.88}$$

together with the closure relation

$$n_\sigma(z_i) = \sum_{occ.} |\varphi_\sigma(z_i)|^2. \tag{3.89}$$

The effective Kohn-Sham potential equation (3.88) includes the extrenal potential $V_{ext}^\sigma(z_i)$, the Hartree potential $V_H^\sigma(z_i)$ and the exchange-correlation potential $V_{xc}^\sigma(z_i)$. For external potential the spin-dependences arise from the strength of confining potentials V_2 in equation (3.51): in principal it can be different for each species. In Section 3.6.3 we will investigate the effect of spin-dependent traps on the ground state densities of spin-up and spin-down species. The spin-dependent Hartree potential can also be written as

$$V_H^\sigma(z_i) = \frac{1}{2} u \, n_{\bar{\sigma}}(z_i), \tag{3.90}$$

where $\bar{\sigma}$ denotes the opposite of spin σ.

However the exchange-correlation contribution needs to be approximated. We notice that exchange interactions between parallel-pseudospin atoms have been effectively eliminated in the Hubbard model by the constraint of allowing double occupation only with opposite spins. Even though we continue to call $E_{xc}[n]$ and $V_{xc}[n]$ the exchange-correlation energy and the exchange-correlation potential, but it is understood that the exchange contribution to these quantities is exactly zero.

The LSDA has been shown to provide an excellent account of the ground-state properties of a large variety of inhomogeneous system [126, 127]. Therefore we employ the following Bethe-Ansatz-based LSDA (BALSDA) functional

$$V_{xc}^{BALSDA} = V_{xc}^{hom}(n, s; u)\Big|_{\substack{n \to n(z_i) \\ s \to s(z_i)}}, \quad (3.91)$$

where, in analogy with *ab initio* DFT, the exchange-correlation potential $V_{xc}^{hom}(n, s; u)$ of the 1D homogeneous Hubbard model is defined by

$$V_{xc\sigma}^{hom}(n, s; u) = \frac{\partial}{\partial n_\sigma}[\varepsilon_{GS}(n, s; u) - \varepsilon_{GS}(n, s; u = 0) - \varepsilon_H]. \quad (3.92)$$

Here $\varepsilon_{GS}(n, s; u)$ is the ground-state energy for the homogeneous case which is provided by the Lieb-Wu equations. The Hartree energy per site ε_H is

$$\varepsilon_H = \frac{1}{2} u \, n_\sigma \, n_{\bar{\sigma}}, \quad (3.93)$$

and $\varepsilon_{GS}(n, s; u = 0)$ is the ground state energy for the noninteracting system

$$\varepsilon_{GS}(n, s; u = 0) = -\frac{4t}{\pi} \sin\left(\frac{\pi}{2}n\right) \cos(\pi s). \quad (3.94)$$

Now we can use the chain rule

$$\frac{\partial}{\partial n_\sigma} = \frac{\partial n}{\partial n_\sigma}\frac{\partial}{\partial n} + \frac{\partial s}{\partial n_\sigma}\frac{\partial}{\partial s}, \quad (3.95)$$

in equation (3.92) and also

$$n = n_\uparrow + n_\downarrow, \quad (3.96)$$

$$s = \frac{n_\uparrow - n_\downarrow}{2}, \quad (3.97)$$

to get a more practical expression for the exchange-correlation potential as a function of chemical potential $\mu(n, s; u)$ and internal magnetic field $h(n, s; u)$:

$$V_{xc\sigma}^{hom}(n, s; u)$$
$$= \mu(n, s; u) \pm \frac{1}{2}h(n, s; u) + 2t \cos\left[\pi(\frac{n}{2} - s)\right] - \frac{1}{2}u \, n_{\bar{\sigma}}. \quad (3.98)$$

The upper (lower) sign refers to the spin-up (spin-down) atoms. Note that the last term here which is the Hartree contribution vanishes in the Kohn-Sham potential equation (3.88), so it does not appear in the effective potential. Illustrative numerical results for the exchange-correlation potential with different values of parameters are reported in Figures 3.5, 3.6 and 3.7. We see that always there is a discontinuity at half-filling $n = 1$ corresponding to the Mott-insulator transition at half-filling.

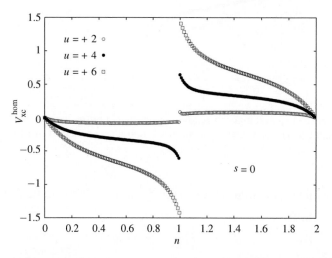

Figure 3.5. The exchange-correlation potential of the 1D homogeneous Hubbard model (in units of t) as a function of the filling n for various values of u.

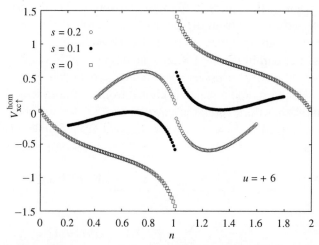

Figure 3.6. The exchange-correlation potential (in units of t) for spin-up component of the 1D homogeneous Hubbard model as a function of filling n.

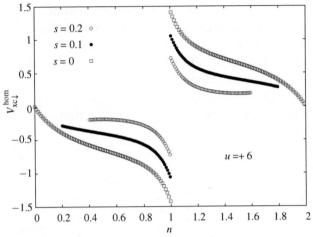

Figure 3.7. The exchange-correlation potential (in units of t) for the spin-down component of 1D homogeneous Hubbard model as a function of filling n.

The chemical potential $\mu(n, s; u)$ and internal magnetic field $h(n, s; u)$ are defined as

$$\mu(n, s; u) = \frac{\partial}{\partial n} \varepsilon_{\mathrm{GS}}(n, s; u),$$

$$(3.99)$$

$$h(n, s; u) = \frac{\partial}{\partial s} \varepsilon_{\mathrm{GS}}(n, s; u).$$

They are the first derivative of the ground-state energy of the homogeneous system with respect to filling n and spin s. These derivatives can be calculated *exactly* from the Bethe-Ansatz ground-state energy. Their exact expressions are given in Appendix (B) and (C).

Therefore, within the the LSDA scheme proposed in equations (3.91) and (3.92), the only necessary input is the xc potential of the 1D homogeneous Hubbard model which is known from its Bethe-Ansatz solution. By generalizing the expression equation (3.27) for the total energy to the spin-resolved case, we can write the total ground-state energy of the system as

$$E_0[n_\uparrow, n_\downarrow] = \sum_{\sigma=\uparrow,\downarrow} \sum_{\alpha} \varepsilon_\alpha^\sigma - \sum_{\sigma=\uparrow,\downarrow} \sum_{i} V_{\mathrm{xc}}^\sigma(z_i)\, n_\sigma(z_i)$$

$$(3.100)$$

$$- \sum_{\sigma=\uparrow,\downarrow} \sum_{i} V_{\mathrm{H}}^\sigma(z_i) + E_{\mathrm{xc}}[n_\uparrow, n_\downarrow].$$

The only term of this energy functional that is needed to be treated within LSDA is the exchange-correlation energy. We can then write the xc energy component via equation (3.92) as

$$E_{xc} \approx \sum_i [\varepsilon_{GS}(n, s; u) + \varepsilon_{GS}(n, s; u = 0) - \varepsilon_H]_{\substack{n \to n(z_i) \\ s \to s(z_i)}} . \qquad (3.101)$$

3.6. Numerical results for repulsive fermions trapped in 1D optical lattices

In this section we present numerical results for the density profiles of a 1D repulsively interacting Fermi gas subject to a harmonic confining potential inside a 1D optical lattice. We first review in Section 3.6.1 recent work by Gao Xianlong et al. [147] on unpolarized gases, highlighting in particular certain ground-state phases which involve metallic and Mott-insulator states. In Section 3.6.2 we then present original numerical results for the spin-resolved atom-density profiles of spin-polarized gases. We study spin-polarized gases in a spin-independent harmonic trap limiting our presentation to metallic phases. In Section 3.6.3 we consider a more general situation of spin-dependent parabolic external potentials. This leads to the possibility of phase separation for sufficiently strong intercomponent repulsion.

Whenever available we will present also numerical results obtained by means of the density-matrix renormalization group (DMRG) technique. These results which have been obtained using the DMRG code released within the "Powder with Power" projec,[1] have been kindly provided by Matteo Rizzi and Prof. R. Fazio.

3.6.1. The unpolarized case: coexistence of metallic and Mott insulating phases

We start the investigation of the 1D repulsive Fermi gas under a harmonic confinement potential equation (3.51) by considering the paramagnetic situation *i.e.*, $N_\uparrow = N_\downarrow$ and spin-independent trap *i.e.*, $V_2^\uparrow = V_2^\downarrow = V_2$. Figure 3.8 shows a sketch of the phases of this gas under harmonic confinement for the case of interspecies repulsions, in dependence of the number of fermions. Starting with a Luttinger liquid in phase $\mathcal{A}$ for $0 < n(z_i) < 1$ one meets

   (i) phase $\mathcal{B}$ where an incompressible Mott insulator occupies the bulk of the trap with the site occupation number $n(z_i)$ locked to 1;

---

[1] www.qti.sns.it

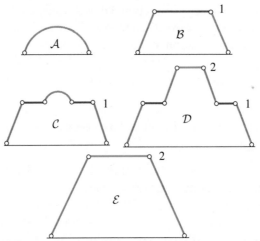

Figure 3.8. Sketch of the phases of a confined Luttinger liquid on a lattice for increasing number of fermions. Adapted from [148].

(ii) phase $\mathcal{C}$ where a fluid with $1 < n(z_i) < 2$ is embedded in a Mott plateau;

(iii) phase $\mathcal{D}$ where a band insulator with $n(z_i)$ locally locked to 2, is surrounded by fluid edges and embedded in a Mott insulator; and finally (iv) phase $\mathcal{E}$ where a band insulator with $n(z_i) = 2$ coexists with fluid edges.

For this paramagnetic system Gao Xianlong *et al.* [147] have solved numerically the self-consistence scheme represented by equation (3.87)-(3.89) by using the local density approximation. Figure 3.9 reports illustrative results for the distribution of site occupation numbers obtained by this DFT theoretical approach in comparison with Quantum Monte Carlo (QMC) method for the metal-insulator phase-separated regime. The agreement of the theory with the simulation data is very remarkable.

3.6.2. Phases of spin-polarized gases

In this section we turn to the spin-imbalance case in which $N_\uparrow \neq N_\downarrow$ under the spin-independent harmonic confinement equation (3.51) with $V_2^\uparrow = V_2^\downarrow = V_2$. The imbalanced spin populations result in a finite spin polarization defined as

$$p = \frac{N_\uparrow - N_\downarrow}{N_\uparrow + N_\downarrow}, \tag{3.102}$$

and the local spin magnetization $m(z_i) = 2\,s(z_i) = n_\uparrow(z_i) - n_\downarrow(z_i)$. We focus our attention to the metallic phases in which $n_\sigma(z_i) < 1$. We have

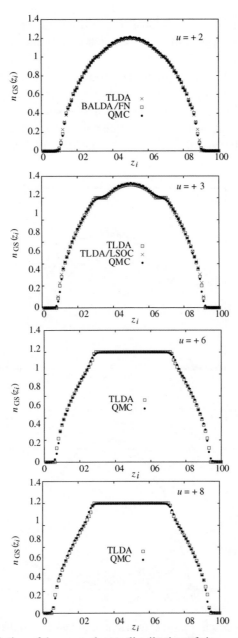

Figure 3.9. Evolution of the ground state distribution of site occupation numbers for $N_\uparrow = N_\downarrow = 70/2$ fermions on a lattice with $N_s = 100$ sites, with increasing interspecies repulsions from top to bottom. These results have been obtained by means of BALDA, QMC, total-energy LDA (TLDA) and the parametrization formula for $\varepsilon_{GS}(n, s; u)$ proposed by Lima *et al.* (LSOC) [149]. From Gao Xianlong *et al.* [147].

solved numerically the self-consistent formalism represented by equations (3.87)-(3.89) for both spin species by using the LSDA in equation (3.91).

In Figures 3.10-3.12 we show the ground-state occupations in a spin-polarized Fermi gas for $p = 0.20$ with the repulsive interaction $u = +2$ to $u = +6$ in a lattice with $L = 100$ lattice sites. L is always chosen so that the trap makes the ground-state spin-resolved site occupation go to zero smoothly near the edges of the lattice. We observe that for all

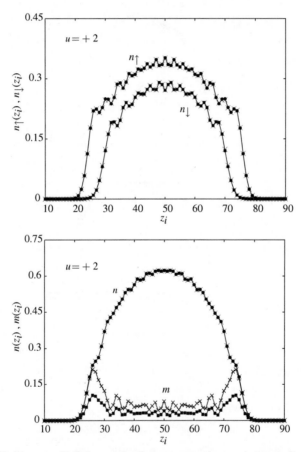

Figure 3.10. Top panel: Spin-resolved site occupation $n_\sigma(z_i)$ as a function of z_i by DMRG (crosses) and BALSDA (filled symbols) for a system with repulsive interaction $u = +2$, $N_\uparrow = 15$ spin-up atoms and $N_\downarrow = 10$ spin-down atoms in $L = 100$ lattice sites, and in the presence of a harmonic potential with $V_2/t = 0.002$. Bottom panel: Total density profiles $n(z_i)$ and local magnetization $m(z_i)$ as a function of z_i for the same system. The thin solid lines are just a guide for the eye.

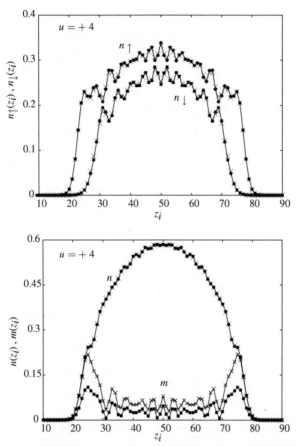

Figure 3.11. Top panel: Spin-resolved site occupation $n_\sigma(z_i)$ as a function of z_i by DMRG (crosses) and BALSDA (filled symbols) for a system with repulsive interaction $u = +4$, $N_\uparrow = 15$ spin-up atoms and $N_\downarrow = 10$ spin-down atoms in $L = 100$ lattice sites, and in the presence of a harmonic potential with $V_2/t = 0.002$. Bottom panel: Total density profiles $n(z_i)$ and local magnetization $m(z_i)$ as a function of z_i for the same system. The thin solid lines are just a guide for the eye.

u spin-up and spin-down densities oscillate out-of-phase. These oscillations cancel each other in the total density and we have a smooth profile for the $n(z_i)$, while the oscillations are enhanced in the local magnetization $m(z_i)$ and there is a spin density wave (SDW) in the bulk of the trap. We have checked with DMRG for very large interaction $u = +50$: in this case the SDW exists in the bulk of the trap. Starting from Figure 3.10 to 3.12 we see that the main effect of the stronger interaction is to broaden the spin-resolved site occupations and also to increase the amplitude of the SDW.

The agreement between the BALSDA and DMRG results is excellent for all values of u. Computationally DMRG is much more expensive than BALSDA. BALSDA takes a few minutes on a small PC to generate a single density profiles and DMRG runs may take about 12 h for a single density profile, but unlike LSDA also provides access to other physical quantities such as the correlation functions.

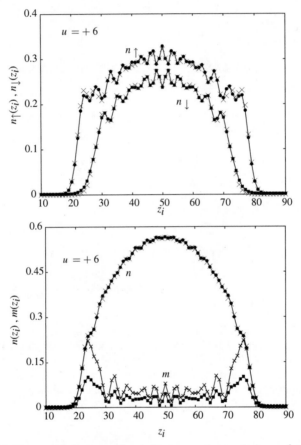

Figure 3.12. Top panel: Spin-resolved site occupation $n_\sigma(z_i)$ as a function of z_i by DMRG (crosses) and BALSDA (filled symbols) for a system with repulsive interaction $u = +6$, $N_\uparrow = 15$ spin-up atoms and $N_\downarrow = 10$ spin-down atoms in $L = 100$ lattice sites, and in the presence of a harmonic potential with $V_2/t = 0.002$. Bottom panel: Total density profiles $n(z_i)$ and local magnetization $m(z_i)$ as a function of z_i for the same system. The thin solid lines are just a guide for the eye.

In Figure 3.13 we show the effect of a different spin polarization p on the spin-resolved density profiles for interaction $u = +2$. In a lattice with

$L = 100$ and a confinement potential $V_2^\uparrow = V_2^\downarrow = 0.002\,t$, we consider a low spin polarization $p = 0.11$ which corresponds to the configuration $(N_\uparrow = 15, N_\downarrow = 12)$ atoms. To get higher polarizations we keep constant the number of spin-up atoms $N_\uparrow = 15$ and we reduce the number of spin-down atoms to $N_\downarrow = 5$ and $N_\downarrow = 3$, corresponding to $p = 0.50$ and $p = 0.66$ respectively. With the fixed spin-up atoms atoms $N_\uparrow = 15$, the spin-up density profile remains the same for different polarization while the spin-down profiles evolves and shows 12, 5 and 3 peaks corresponding to $N_\downarrow = 12, 5$ and 3 atoms with spin-down. For simplicity Figure 3.13 just shows the total density (top panel) and the local magnetization (bottom panel) for different polarizations.

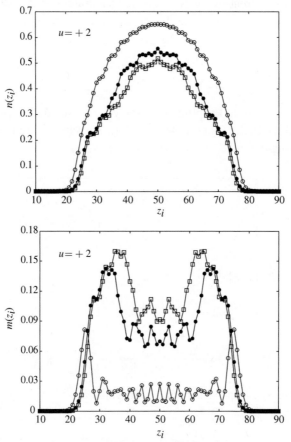

Figure 3.13. Evolution of the ground state distribution of the total ground-state population (top panel) and the local magnetization (bottom panel) on a lattice with $L = 100$ sites, with increasing spin polarization. In every panel polarization $p = 0.11$ is denoted by circle, $p = 0.50$ by filled circle and $p = 0.76$ by square. The thin solid lines are just a guide for the eye.

We see with the spin imbalanced populations in the spin-independent trap, always both species are in the bulk of the trap: no phase separation occurs. In the next section we take a spin-dependent trap, which leads to phase separation between spin-up and spin-down atoms in the bulk of the trap.

3.6.3. Phase separation in spin-dependent external potentials

In this section we generalize the previous study of a spin-independent trap to the spin-dependent case. The spin dependence in the confining potential arises from different strengths $V_2^\uparrow \neq V_2^\downarrow$. The spin-resolved Kohn-Sham equations (3.87)-(3.89) and exchange-correlation potential (3.98) allow to treat every spin species separately with a different trap strength. We quantify the ratio between two trap strength as

$$\lambda = \frac{V_2^\uparrow}{V_2^\downarrow}. \tag{3.103}$$

We increase λ while keeping constant $V_2^\downarrow$ i.e. we make spin-up atoms more confined in the center of the trap. As an example Figure 3.14 shows the confining potentials for the case $\lambda = 3$.

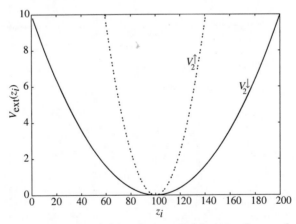

Figure 3.14. Spin-dependent harmonic potential in unit of t as a function of z_i for $\lambda = 3$ and $V_2^\downarrow/t = 10^{-4}$.

To see the effect of interplay between the spin-dependent harmonic trap and the intercomponent repulsive interaction on the spin-polarized Fermi gas, we consider an asymmetric mixture of $N_\uparrow = 30$ spin-up atoms and $N_\downarrow = 10$ spin-down atoms in a lattice with $L = 200$ sites. We compare the results of density profile and local magnetization for weak interaction $u = +1$ and strong interaction $u = +4$. For three different values of $\lambda = 1$, 3 and 6 we show in the Figure 3.15 the spin-resolved density profiles of the mixture.

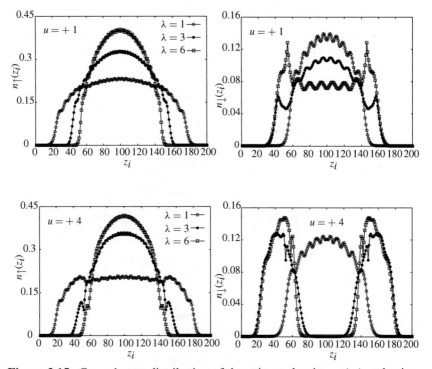

Figure 3.15. Ground-state distribution of the spin-up density $n_\uparrow(z_i)$ and spin-down density $n_\downarrow(z_i)$ in an axially confined lattice-Hubbard gas with the weak repulsive interaction $u = +1$ (top panels) and strong repulsive interaction $u = +4$ (bottom panels) for increasing relative trap strength λ. The gas consists of $N_\uparrow = 30$ and $N_\downarrow = 10$ fermions on a lattice of $L = 200$ sites. In all cases the spin-down trap strength is equal to $V_2^\downarrow/t = 10^{-4}$. The thin solid lines are just a guide for the eye.

For the weak interaction $u = +1$ (top panels), upon increasing λ we observe more confined spin-up atoms in the bulk of the trap where, because of the repulsive interaction, a depletion occurs in the spin-down density. However the spin-down atoms still exist in the bulk and the weak interaction can not push them to the edges of the trap. The bottom panels in Figure 3.15 show density profiles in the case of strong repulsive interaction $u = +4$. In this regime and for $\lambda > 1$ the strong repulsive interaction prohibits the existence of spin-down atoms in the bulk and they have to move to the periphery of the trap. Therefore we see clearly there is complete *phase-separation* between two components in the bulk of the tarp.

Figure 3.16 shows also this fact in the behavior of the local magnetization $m(z_i)$. For $\lambda > 1$ in the periphery there is large accumulation of spin-down atoms therefore the local magnetization becomes more and more negative with increasing the strength of the interaction from weak $u = +1$ to the strong interaction $u = +4$. Instead in the bulk of the trap the local magnetization increase because of the larger number of spin-up atoms.

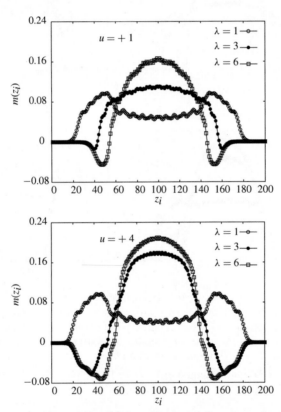

Figure 3.16. Ground-state distribution of the on-site magnetization $m(z_i)$ in an axially confined lattice-Hubbard gas with repulsive interaction $u = +1$ (top panel) and $u = +4$ (bottom panel), for increasing relative trap strength λ. The gas consists of $N_\uparrow = 30$ and $N_\downarrow = 10$ fermions on a lattice of $L = 200$ sites. In all case the spin-down trap strength is equal to $V_2^\downarrow/t = 10^{-4}$. The thin solid lines are just a guide for the eye.

Chapter 4
Ultracold attractive fermions in one-dimensional optical lattices

In this chapter we deal with Fermi gases with attractive interactions subject to parabolic trapping inside a 1D optical lattice. We first review the theory of various exotic bulk superfluid phases such as the Larkin-Ovchinnikov-Fulde-Ferrell (LOFF) phase, the breached-pair superfluid state and phase separation in fermion superfluids. These phases result from pairing fermions of different species with unequal densities. We mention the experimental realization of this kind of Fermi gas carried out at MIT and at Rice University. Turning to 1D systems, Gao Xianlong *et al.* [150] have analyzed unpolarized fermions with attractive interaction confined by a parabolic trap. They have shown the emergence of atomic density waves induced by harmonic confinement. We have generalized the calculations on unpolarized fermionic system by allowing a finite spin polarization *i.e.* imbalanced spin populations. We observe that a sizeable fraction of majority-spin atoms accumulates at the edges of the trap, leaving a core of paired atoms at the center of the trap. This "spin-separated" ground state will be investigated by solving the mean-field Bogoliubov-de Gennes (BdG) equations. The BdG equations provide information on both the spin-resolved density profiles and the local pairing gap. We observe that the local pairing gap (i) is in the unpolarized case a flat function of position in the bulk of the trap; (ii) oscillates at the edges of the trap as soon as a spin imbalanced introduced, while remaining flat in the bulk of the trap; and (iii) becomes a highly-oscillating function everywhere in the trap with increasing spin-polarization, eventually approaching zero in the fully-spin polarized case.

4.1. Exotic superfluid states in condensed matter physics

The formation of electron pairs, each pair consisting of a spin-up and a spin-down electron, underlies the phenomenon of s-wave superconductivity. While the population of each spin component is generally equal in superconductors, an imbalance is produced in experiments with gases of trapped, ultracold fermionic atoms. Exotic new states of matter are predicted for the imbalanced system that, if realized, may have important implications for our understanding of nuclei, compact stars, and quantum chromodynamics. In this section we review some of these exotic phases.

4.1.1. The LOFF phase

Some 40 years ago Larkin and Ovchinnikov (LO) [151] and Fulde and Ferrell (FF) [152] proposed on purely theoretical grounds what amounted to a new type of superconducting state, now often referred to as the LOFF phase or as inhomogeneous superconductivity. This can arise when the species participating in the pairing phenomenon have different Fermi surfaces with a large enough separation. In this state the superfluid condensate varies in space.

The LOFF phase can be viewed as a further generalization of the BCS state: whereas the building block of the BCS theory is the Cooper pair, where the two electrons have momenta equal in magnitude and opposite in direction, in the LOFF phase a salient feature is that momenta do not add to zero. Then an almost immediate consequence of the LOFF proposal is that the energy gap, i.e. the order parameter of the superconducting state, varies in space: the ground state is inhomogeneous, and even crystalline structure might be formed [153, 154]. This proposal has not yet been confirmed beyond reasonable doubt in condensed matter: its observation would necessitate the employment of high magnetic fields and type-II superconductors which should be essentially devoid of impurities. Studies of organic and heavy-fermion materials or layered superconductors with magnetic fields parallel to the layers, and observations on surface superconductivity, vortex lattices, and Josephson junctions between LOFF and BCS materials have all been addressed in the recent literature [155].

The basic ideas of LOFF may also prove to play an important role in the future in nuclear physics and in the theory of some aspects of the properties of pulsars as commonly identified with neutron stars [153, 154]:

(i) *Nuclear physics.* Neutron-proton (n-p) correlations and the likelihood of n-p Cooper pair condensation are at present being studied in a number of different contexts, including the fields of heavy ion collisions and of quark matter. The possibility of a spatially non-homogeneous condensate in asymmetric nuclear matter has been examined.

(ii) *Astrophysics.* It is presently conjectured that inhomogeneous superconductivity could be generated by the difference in quark chemical potentials brought about by weak interactions in the inner core of pulsars. This might well afford a mechanism for explaining glitches in pulsars.

Let us recall the argument of FF [152]. These authors set out to examine a specific problem concerning magnetic impurities in a non-magnetic metal in the case when the host metal becomes superconducting. However, the problem of applying a magnetic field on a superconductor has already been considered a long time ago. Here this is the applied magnetic field which tends to make two spin populations unequal. Usually the critical field is limited by orbital effects. However the question of limitation of the superconducting phase, when these orbital effects are small, has been considered very early and it was pointed out by Clogston [156] that the standard BCS phase could, at most, resist to a difference in chemical potential between the two populations of the order of the critical temperature (the so-called Pauli paramagnetic limit). Not long after, Fulde and Ferrell showed that the pairing could somewhat adjust to the difference in chemical potential, instead of just resisting, by letting the pairs have a common momentum $\mathbf{k}$ instead of having it equal to zero as for a standard (BCS) superconductor. However the effect was actually found to be rather small. Indeed at zero temperature the standard BCS phase goes to the normal state by a first-order transition [156] when the chemical potential difference $2\delta\mu$ between chemical potential of the $\mu_\uparrow = \mu + \delta\mu$ and $\mu_\downarrow = \mu - \delta\mu$ is equal to $\delta\mu_1 = \frac{\Delta_0}{\sqrt{2}}$, where $\Delta_0 = 1.76\,T_c$ is the zero-temperature gap for equal populations. The LOFF phase goes to the normal state by a second-order transition for $\delta\mu_2 = 0.754\,\Delta_0$, which is not much beyond [157]. Therefore the anisotropic superconducting phase can exist only in a narrow window [153]

$$\delta\mu_1 < \delta\mu < \delta\mu_2. \tag{4.1}$$

At the critical point one can view the appearance of the LOFF phase in the following way. When the chemical potentials for the two spin populations are different, it becomes more costly in terms of kinetic energy to form $(\mathbf{k}, -\mathbf{k})$ pairs in standard BCS way, because one cannot pick the two particles very near the Fermi surface, since these surfaces do not match due to their size difference [157]. In the LOFF phase this problem is remedied by taking a nonzero total momentum so that the paired particles are both near their respective Fermi surfaces. According to LOFF the pairs have momenta

$$\mathbf{p}_\uparrow = \mathbf{k} + \mathbf{q}, \qquad \mathbf{p}_\downarrow = -\mathbf{k} + \mathbf{q} \tag{4.2}$$

so that $\mathbf{k}$ identifies a particular pairs, and every pair in the condensate has the same nonzero total momentum $2\mathbf{q}$ [154]. Since the total momen-

tum of the pair is not zero the condensate breaks rotational and transla-
tional invariance, the simplest form of the condensate compatible with
this breaking is just a plane wave [153]

$$\Delta(\mathbf{r}) = \Delta_0 \, e^{2i\mathbf{q}\cdot\mathbf{r}} . \tag{4.3}$$

It should also be noticed that the pairs use much less of the Fermi surface
that they do in the BCS case. More generally there is a quite large region
in momentum space (the so-called blocking region) which is excluded
from the pairing. The effect of the blocking region is to reduce the phase
space where pairing is possible. This leads to a condensate smaller than
the BCS one. The complementary phase space, where pairing is possible,
will therefore be called the pairing region [153].

The essential point of the LOFF proposal is illustrated in Figure 4.1.
The top panel shows type-S (single) depairing in momentum space as
produced by the spin-exchange field in conjunction with a shift of the
Fermi sphere to the right. The shaded portion in (a) is occupied by down-
spin electrons which are stabilized by the exchange field, and the shaded
region in (b) is empty of spin-up electrons. The remaining regions of momentum space are available for pairing, but the magnitude of the energy

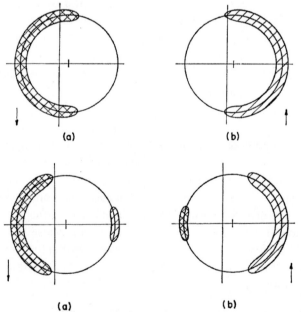

Figure 4.1. Illustrating S-type pairing (top) and D-type pairing (bottom) in the
FF proposal. Adapted from [152].

gap is reduced because of the reduction in the available phase space. The bottom panel in Figure 4.1 illustrates type-D (double) depairing: the shift of the Fermi surface to the right is sufficiently large that also some part of unpaired spin-up electrons is stabilized in (a), leaving a vacant region for down-spin electrons in (b).

4.1.2. Breached-pair superfluidity

The breached-pair (BP) state, proposed by Liu and Wilczek [158] is a non-BCS superfluid phase with gapless fermionic excitations. A unique features of this state is its phase separation in momentum space, although in real space it is a homogeneous, polarized superfluid. This feature distinguishes the BP phase from a single mixture of molecular condensate and fermionic atoms. Stimulated by questions in quantum chromodynamics (QCD) at high density [154], Liu and Wilczek proposed a new state of matter, the so-called interior-gap superconductivity in which the pairing carves out a gap within the interior of a large Fermi sphere while the exterior surface remains gapless.

These authors considered a homogeneous fermion gas in 3D containing two species ($\alpha = A, B$) obeying parabolic dispersion relations, described by the Hamiltonian

$$\mathcal{H} = \sum_{\mathbf{p}\alpha} \varepsilon_\alpha(\mathbf{p}) \, \psi_\alpha^\dagger(\mathbf{p})\psi_\alpha(\mathbf{p})$$

$$+ g \sum_{\mathbf{pp'q}} \psi_A^\dagger(\mathbf{q}+\mathbf{p}) \, \psi_B^\dagger(\mathbf{q}-\mathbf{p}) \, \psi_B(\mathbf{q}+\mathbf{p'}) \, \psi_A(\mathbf{q}+\mathbf{p'}) \tag{4.4}$$

where $\varepsilon_\alpha(\mathbf{p}) = \mathbf{p}^2/2m_\alpha - \mu_\alpha$ and $g = -4\pi\hbar^2 a_s/\tilde{m}$ where a_s is the scattering length for s-wave scattering between species A and B, and $\tilde{m}$ is the reduced mass. The chemical potentials are defined so that the Fermi surface for both species are at $\varepsilon_F = 0$. Being in a weak coupling regime, the ground state can be constructed by modifying the ground state of the noninteracting system. If the Fermi surfaces matched, the attractive interaction would trigger standard BCS superfluidity. The BCS wave function, however, postulates either zero or double occupied paired modes and it is incompatible with keeping the modes of species B between p_F^A and p_F^B completely filled. To support pairing of total momentum zero it is needed to promote some particles of species B up to momentum near p_F^B, thus carving an interior "gap" of the species B Fermi sea near momentum $p = p_F^A$.

The Liu and Wilczek calculations show that the gap equation supports a nonzero solution only for $|g| > g_c$, with

$$g_c \simeq 2 \left[N_+(0) \ln \left(\frac{p_0 \lambda}{p_F^{B\,2} - p_F^{A\,2}} \frac{m_A + m_B}{m_A} \right) \right]^{-1}, \qquad (4.5)$$

where the generalized density of states $N_+(0)$ has been introduced as $N_+(0) \equiv \sum_\mathbf{p} \delta(\varepsilon^+(\mathbf{p}))$ and p_0 is the point where $\varepsilon^+(\mathbf{p}) = 0$, with $\varepsilon^+(\mathbf{p}) \equiv \frac{1}{2}(\varepsilon^A(\mathbf{p}) + \varepsilon^B(\mathbf{p}))$. However, $g_c \to 0$ when $m_B/m_A \to \infty$ for fixed $p_F^{A,B}$, so that interior gap superfluidity can be favorable for arbitrarily weak attractive interaction.

This theory has been criticized by questions of Bedaque *et al.* [159] on instability toward phase separation and by Wu and Yip [160] on local current. Forbes *et al.* [161] clarified and corrected the discussion of stability criteria. They concluded that: (i) For extensive systems, one cannot stabilize a state by imposing a global constraint (such as fixed particle numbers): the composition of the state can be completely determined from an analysis in the grand canonical ensemble; (ii) with the proper momentum structure, however, one may realize BP superfluidity in states that are thermodynamically stable for fixed chemical potentials.

4.1.3. Phase separation in fermion superfluids

Here we illustrate this type of studies by reporting the analysis given by Bedaque *et al.* for phase separation in homogeneous fermion superfluids. These authors consider a dilute gas of two species A and B of particles, with chemical potentials $\mu_{A,B}$ and masses $M_{A,B}$, and start from a pairing Hamiltonian in the form

$$\begin{aligned}
\mathcal{H} - \sum_{i=A,B} \mu_i N_i &= \sum_\mathbf{k} \varepsilon_\mathbf{k}^{(i)} \hat{\psi}_i^\dagger(\mathbf{k}) \hat{\psi}_i(\mathbf{k}) \\
&+ g \sum_{\mathbf{k},\mathbf{k}'} \hat{\psi}_A^\dagger(\mathbf{k}') \hat{\psi}_B^\dagger(-\mathbf{k}') \hat{\psi}_B(-\mathbf{k}) \hat{\psi}_A(-\mathbf{k}),
\end{aligned} \qquad (4.6)$$

with dispersion relation $\varepsilon_\mathbf{k}^{(i)} = \mathbf{k}^2/(2M_i) - \mu_i$. A mean-field treatment, with the definition $E_\mathbf{k}^{(\alpha,\beta)} = \pm \varepsilon_\mathbf{k}^{(-)} + \sqrt{[\varepsilon_\mathbf{k}^{(+)}]^2 + \Delta^2}$ where $\varepsilon_\mathbf{k}^{(\pm)} = [\varepsilon_\mathbf{k}^{(A)} \pm \varepsilon_\mathbf{k}^{(B)}]/2$ and $\Delta = -g \sum_\mathbf{k} \langle \hat{\psi}_B^\dagger(-\mathbf{k}) \hat{\psi}_A^\dagger(\mathbf{k}) \rangle$, leads to a BCS-like state $[u_\mathbf{k} + v_\mathbf{k} \hat{\psi}_B^\dagger(-\mathbf{k}) \hat{\psi}_A^\dagger(\mathbf{k})]$ in the modes $\mathbf{k}$ where $E_\mathbf{k}^{(\alpha,\beta)} > 0$, but to a state filled with particle B (for the case $M_B > M_A$) in the modes where $E_\mathbf{k}^{(\beta)} < 0$. For some value of Δ the energy $E_\mathbf{k}^\beta$ may be negative in the range of momentum $k_1 \le k \le k_2$ where

$$k_{1,2}^2 = \frac{1}{2} \left[p_A^2 + p_B^2 \pm \sqrt{(p_B^2 - p_A^2) - 16 M_A M_B \Delta^2} \right], \qquad (4.7)$$

with $p_i = \sqrt{2M_i\mu_i}$ being the Fermi momenta of the two species. The thermodynamic potential is given by

$$\Omega = \langle \mathcal{H} - \sum_{i=A,B} \mu_i N_i \rangle$$

$$= -\frac{M\Delta^2}{2\pi a_s} + \sum_{\mathbf{k}} \left[\Theta(-E_{\mathbf{k}}^\beta) E_{\mathbf{k}}^\beta + \varepsilon_{\mathbf{k}}^{(\beta)} - E_{\mathbf{k}}^\beta \right] \tag{4.8}$$

where M is the reduced mass and a_s is the interspecies scattering length $(g = 2\pi a_s/M)$.

The results of a stability analysis depend on whether one works at fixed chemical potentials or at fixed particle densities.

(i) *Fixed chemical potentials.* Starting with a BCS ground state for $\mu_A = \mu_B$ and increasing $p_B^2 - p_A^2$ at fixed $p_0^2 = M(p_B^2/M_B - p_A^2/M_A)$, one finds that for $\Delta < (p_B^2 - p_A^2)/(4\sqrt{M_A B})$ the thermodynamic potential can be lowered by filling with B particles the states in the range given by equation 4.7: therefore, there is a first-order phase transition between the superfluid and the normal state. The interior-gap state corresponds in this case to a maximum of the thermodynamic potential. Figure 4.2 shows the thermodynamic potential Ω as a function of Δ computed from a numerical evaluation of equation (4.8).

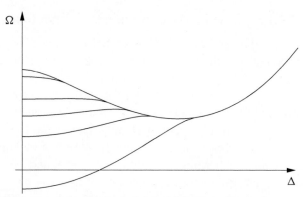

Figure 4.2. Thermodynamic potential for different values of p_B and p_A (constant p_0). The top curve corresponds to $p_A = p_B$ and the lower curves correspond to increasing values of $|p_B^2 - p_A^2|$. Adapted from [159].

(ii) *Fixed partial densities.* Whereas in the BCS state the particle numbers are constrained to be the same for the two species, one may have $n_B \neq n_A$ in both the normal state and in the interior-gap state. The latter would simultaneously be a superfluid and a Fermi liquid with two

Fermi surfaces corresponding to k_1 and k_2. Bedaque *et al.* write the energy of a mixture of normal and BCS phases as a weighted average of their energies, given for small values of the gap as

$$
\begin{cases}
E_N(n_A, n_B) = \dfrac{(6\pi^2 n_A)^{5/3}}{20\pi^2 M_A} + \dfrac{(6\pi^2 n_B)^{5/3}}{20\pi^2 M_B} \\
E_{BCS}(n_A = n_B = n) = \dfrac{(6\pi^2 n)^{5/3}}{20\pi^2 M} - \dfrac{(6\pi^2 n)^{1/3} M \Delta^2(n)}{2\pi^2}.
\end{cases}
\tag{4.9}
$$

They conclude that within mean-field theory, for all reasonable values of the parameters at fixed n_A and n_B, a mixed state consisting of bubbles of the normal phase in a sea of the BCS phase is lower in energy than the interior-gap state.

(iii) *Fixed total density.* A third case is finally considered, in which the total density is kept fixed but conversion of particles from one species to the other is allowed. This would be relevant in a relativistic formulation for the physics of high-density quark matter, where weak interactions can change the flavor of the quarks. From their non-relativistic formulation Bedaque *et al.* find again that a mixture of normal and superfluid parts is the energetically favoured state at $\mu_A = \mu_B$, becoming pure BCS for $M_A = M_B$ and pure normal B-species when $M_B \gg M_A$.

4.2. The LOFF phase in 1D spin-polarized fermionic systems

As discussed in Chapter 1 (see Section 1.1.5), a system of attractive unpolarized fermions is in the Luther-Emery phase. In this section we illustrate an analogue of the LOFF in 1D spin-polarized attractive fermions.

Machida *et al.* [162] have analyzed the possibility that a spatially modulated superfluid LOFF state could be observed in fermion atom condensates. They start out from a pairing Hamiltonian of the type shown in equation (4.6) with interspecies attractions ($g < 0$), but including in addition an axisymmetric harmonic potential with radial and axial frequencies ω_r and ω_z, and use chemical potentials of the type $\mu_\uparrow = \mu + \delta\mu$ and $\mu_\downarrow = \mu - \delta\mu$ to prepare unequal populations of the two species. The Hamiltonian is diagonalized by a Bogoliubov transformation, leading to Bogoliubov-de Gennes equations with a self-consistent condition for the determination of a position-dependent gap parameter $\Delta(\mathbf{r})$. Analytic results are first obtained in the limit of a long cigar-shaped gas ($\omega_z/\omega_r \to 0$), corresponding to a quasi-1D system in the absence of a trap. They find in this limit [163]

$$
\Delta(z) = \Delta_1 k_1 \, \text{sn}(\Delta_1 z, k_1)
\tag{4.10}
$$

where $\mathrm{sn}(z, k)$ is a Jacobi elliptic function and Δ_1 is the order parameter to be determined self-consistently. From this solution they find that the LOFF state is energetically stable when the relative population difference satisfies the inequality

$$\frac{|n_\uparrow - n_\downarrow|}{n_\uparrow + n_\downarrow} \geq \frac{\Delta_0}{E_\mathrm{F}} \tag{4.11}$$

where Δ_0 is the order parameter for a uniform BCS state at zero temperature. Beyond this critical population imbalance the uniform BCS state changes into the modulated LOFF state, with the excess density of spin-up fermions accumulating periodically at the zeros of $\Delta(z)$.

Machida *et al.* suggest that the the spin-density modulation may be revealed by separate optical imaging of each atomic species or by means of Stern-Gerlach experiments. As a rough estimate they give its wavelength as $3 - 30 \, \mu\mathrm{m}$ for $\Delta_0/E_\mathrm{F} = 0.1 - 0.01$ at a particle density of 10^{20} m^{-3}. Their conclusions are then supported by detailed numerical calculations on a trapped gas, leading to illustrations of the spatial profiles of the pairing field $\Delta(r, z)$ and of the spin density $\rho(r, z)$.

Yang [164] has treated the inhomogeneous superfluid state in a system of weakly coupled chain and found a phase diagram in which the system goes from the uniform to the non-uniform state through a continuous transition of the commensurate-incommensurate type. We focus here on his results for a single chain, that he treats by an exact bosonization method in contrast to the mean-field treatment of Machida *et al.* . His basic Hamiltonian for the 1D system contains, in addition to terms having the form of equations (1.78)-(1.80) to describe the independent charge and spin sectors in a Luther-Emery liquid, a supplementary term to represent a Zeeman coupling of the form $H_z = (2\pi)^{-1}\tilde{\mu}B \int dx \partial_x \phi_S(x)$ ($\tilde{\mu}$ being the magnetic moment of a particle). With reference to the discussion in Section 1.1.5, the charge sector is described by free massless bosons, whereas the spin excitations are massive solitons carrying spin $\pm 1/2$. The spin gap Δ_s is in a sense the analog of the BCS gap in higher-dimensional superconductors, except that in 1D there cannot be long-range superconducting order but the correlation function of the Cooper-pair operator must decay with a power law. The Zeeman field couples to the spin-soliton density and plays the role of the chemical potential for the spin solitons. The expectation is that a continuous transition occurs at a critical field given by

$$B_c = 2\Delta_\mathrm{s}/\tilde{\mu}, \tag{4.12}$$

beyond which spin solitons start to proliferate in the ground state. The new phase in the spin sector is the 1D analog of the LOFF phase in higher-dimensional systems, since the appearance of spin solitons in the ground state induces an oscillatory phase in the superfluid correlation function.

In a comparison with the 1D mean-field treatment of Machida *et al.*, Yang points out [165] that his exact solution for a strictly 1D system, which takes account of quantum fluctuations, suggests a significantly wider parameter range for the LOFF state. Instead of obeying the inequality given in equation (4.11), the population difference δn increases continuously in proportion to $\sqrt{B - B_c}$. While only power-law order is possible in 1D, a weak 3D coupling will stabilize true long-range order and yield a spatially oscillating order parameter in the LOFF phase with wave number $q \propto n$.

Machida *et al.* [162] also discuss the possibility of realizing the LOFF state by measuring the modulation in local magnetization, which reflect the underlying structure of the pairing order parameter. Yang proposed [165] two alternative methods to detect the LOFF state, which directly probe the momenta of the Cooper pairs. The first method relies on the time-of-flight measurement to determine the molecular velocity distribution. One can do the experiment of time-of-flight in the BEC-BCS crossover [76] on the LOFF state; the fundamental difference is that in this case because the Cooper pairs carry intrinsic (non-zero) momenta, and the condensate will show up as peaks corresponding to a set of *finite* velocities in the distribution. Another method to detect the Cooper pairs is to study the correlation in the shot noise of the fermion absorption images in time-of-flight, first proposed by Altman *et al.* [166]. Greiner *et al.* [167] showed that the noise correlation clearly demonstrates correlation in the occupation of **k** and -**k** states in momentum space when weakly bound diatoms molecules are dissociated and the trap is removed. In principle the same measurement can be performed on fermionic superfluid states, and for the LOFF state, it would reveal correlation in the occupation of **k+q** and $-\mathbf{k} + \mathbf{q}$ states, where **q** is one of the momenta of the pairing order parameter. Both methods allow one to directly measure **q**, which defines the LOFF state. These method are unique to the cold atom systems; in superconductors the only comparable method is Josephson effect [168].

4.3. Experimental results on superfluid atomic Fermi gases with imbalanced spin populations

Recently experiments studying spin-polarized Fermi gases were done by Zwierlein *et al.* [169] at MIT and by Partridge *et al.* [170] at Rice Univ. Ketterle's group has explored also the imbalanced Fermi gas by direct observation of the superfluid phase transition [171] and by observation of phase separation [172]. In this section we review the experiments carried out in Ketterle's group.

4.3.1. Superfluid-to-normal quantum phase transition

One of the main difficulties in stabilizing non-conventional supercon- ducting states in the condensed matter context lies in the fact that ap- plying a field to create a population difference of up and down electrons inevitably induces the diamagnetic current which acts as a Cooper-pair breaker. The present neutral cold fermion systems are ideal for pursuing the realization of the exotic phases *e.g.* the LOFF state: (1) There are no pair breaking mechanisms such as the diamagnetic current or impurities which weaken the stability of this state. (2) Fine adjustment on the popu- lation difference of two species and attractive interactions is possible for the system to be in the most favorable and easily observable condition for the LOFF state [162]. Population imbalance in cold-atom gases plays essentially the same role as the Zeeman or exchange field in supercon- ductors, since pairing is dependent on energy measured from the Fermi energy for each species of fermions [173].

The discovery of fermion superfluids [174] provides a new possibility of exploring unequal mixtures of fermions, because populations in two hyperfine states of the fermionic atoms can be chosen freely. In equal mixtures of fermions, a Feshbach resonance gives access to the BEC- BCS crossover. In the case of unequal mixtures, studied by Zwierlein *et al.* [169], all fermions in the less populated spin state will pair up with atoms in the other state. The resulting condensate will spatially coexist with the remaining Fermi sea of unpaired atoms. As the repulsive interac- tion between atoms and molecules is increased, the condensate will start to expel unpaired atoms, leading to phase separation of the superfluid from the normal phase [175].

In the BCS side of the Feshbach resonance, the pairing gap Δ pre- vents unpaired atoms from entering the BCS superfluid [159, 175]. As the binding energy and hence the pairing gap is further reduced, Δ will eventually become small compared to the chemical potential difference $\delta\mu$, allowing unpaired excess atoms to enter the superfluid region. Close to this point, superfluidity will cease to exist. In the weakly interacting BCS limit, the pairing gap is exponentially small compared to the Fermi energy,

$$\Delta = E_{\mathrm{F}} \exp\left(-\frac{\pi}{2k_{\mathrm{F}}|a|}\right) \qquad (4.13)$$

and hence an exponentially small population imbalance can destroy superfluidity. This quantum phase transition from superfluid-to-normal state can be driven by increasing the mismatch in chemical potentials between the two spin states to the critical value of $\delta\mu \approx \Delta$, inducing collapse into the normal state. This type of phase transition corresponds to Pauli (Clogston) limit of superconductivity.

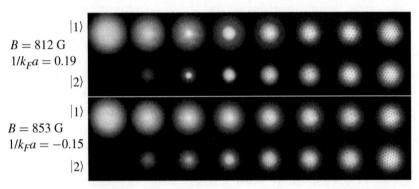

Figure 4.3. Superfluidity in a strongly interacting Fermi gas with imbalanced spin populations. The upper (lower) pair of rows shows clouds prepared at the BEC side (BCS side). In each pair of rows, the upper image shows state $|1\rangle$, the lower one state $|2\rangle$. For the 812-G data, the population imbalance $p = (N_2 - N_1)/(N_2 + N_1)$ between N_1 atoms in state $|1\rangle$ and N_2 in state $|2\rangle$ was (from left to right) 100, 90, 80, 62, 28, 18, 10, and 0%. For the 853-G data, the imbalance was 100, 74, 58, 48, 32, 16, 7, and 0%. Adapted from Zwierlein *et al.* [169].

Figure 4.3 shows images of the two spin states for varying population imbalance on both sides of the resonance. Starting with a pure Fermi sea in state $|1\rangle$, it can be seen how gradually, for increasing number in the second spin state $|2\rangle$, first a normal (uncondensed) cloud of fermion pairs emerges, then a condensate peak appears within the normal clouds (Figure 4.4, a and b). As the condensate size increase and the friction due to the normal component decreases, vortices appear in the rotating cloud, a direct signature of superfluid flow. As expected, the largest condensate with the largest number of vortices are obtained for an equal mixture (Figure 4.5). However, superfluidity in the strongly interacting Fermi gas is clearly not constrained to a narrow region around the perfectly balanced spin mixture, but is observed for large population asymmetries.

Figures 4.3 and 4.5 summarize the results for rotating spin mixtures and displays the number of detected vortices versus the population imbalance between the two spin states. The vortex number measures qualitatively how deep the system is in the superfluid phase. Figure 4.5 therefore shows the shrinking of superfluid region with decreasing interaction strength on the BCS side, closing in on the optimal situation of equal populations.

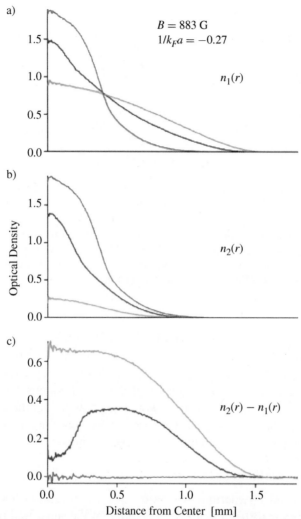

Figure 4.4. Radial density profiles of the two components of a strongly inter-acting Fermi-gas mixture with unequal populations. The profiles are azimuthal averages of the axially integrated density. The imaging procedure involves a magnetic-field sweep and ballistic expansion. The population imbalance was $p = 0\%$ (red), $p = 46\%$ (blue), and $p = 86\%$ (green). The clear dip in the blue curve in (c) caused by the pair condensate indicates phase separation of the superfluid from the normal gas. From Zwierlein *et al.* [169].

A property of the superfluid state with imbalanced populations is the clear depletion in the excess fermions of the majority component (Figure 4.4 c). The profiles in Figure 4.4 present the axially integrated density; hence,

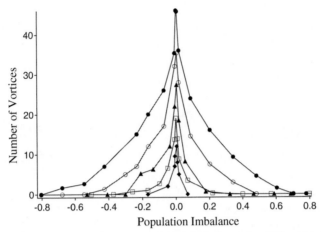

Figure 4.5. Vortex number versus population imbalance for different interaction strengths. Results are shown for 812 G ($1/k_Fa = 0.2$, •), 853 G ($1/k_Fa = -0.15$, o), 874 G ($1/k_Fa = -0.3$, ▲), 896 G ($1/k_Fa = -0.4$, □), and 917 G ($1/k_Fa = -0.5$, ◆). From Zwierlein *et al.* [169].

the true depletion in the 3D density is even stronger. The condensate seems to repel the excess fermions. Therefore, the depletion observed in expansion hints at spatial *phase separation* of the superfluid from the normal cloud. However Zwierlein *et al.* [169] did not observed a modulation in the condensate density as would be predicted for the LOFF state [162, 176]. This state is predicted to be favored only in a narrow region of parameter space and might have escaped attention.

4.3.2. Observation of phase separation

In the successive experiments with unequal mixture of Fermi gas, Ketterle's group reported the direct observation of the superfluid phase transition in a strongly interacting gas of ^{6}Li fermions [171] and the observation of phase separation [172]. Zwierlein *et al.* [171] demonstrated that the *normal-to-superfluid* phase transition in a strongly interacting Fermi gas can be directly observed in absorption profiles. As for BEC, the phase transition is observed as a sudden change in the shape of the cloud during time-of-flight expansion, when the trap depth (corresponds to the temperature) is decreased below a critical value. Figure 4.6 shows schematically this change.

Figure 4.7 shows column density profiles of the two imbalanced spin states for different points along the evaporation path corresponding to different temperatures, and for three magnetic fields that correspond to the the BEC side, to the resonance, and to the BCS side of the resonance. For

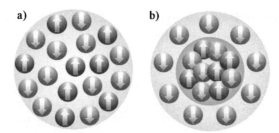

Figure 4.6. a) A homogeneous state of an unequal number of spin-up and spin-down atoms, such as that prepared by Zwierlein *et al.* [171]. b) As the gas temperature is lowered past a critical value, a phase transition occurs. A superfluid state forms in the center of the gas with spin-up and spin-down atoms paired and equal in number, and the excess unpaired atoms move to outside the central core. Adapted from Thomas [177].

large final trap depths (upper panels) the smaller cloud has the expected shape of a normal, non-superfluid gas: it is very well fitted using a single,

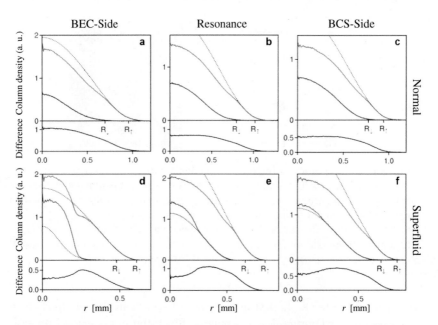

Figure 4.7. Direct observation of the phase transition in a strongly interacting two-state mixture of fermions with imbalanced spin populations. Top (**a-c**) and bottom (**d-f**) rows show the normal and the superfluid state, respectively. The dotted lines show Thomas-Fermi fits to the wings of the column density. The radii $R_\uparrow$ and $R_\downarrow$ mark the Fermi radius of an expanding, non-interacting cloud with atom number $N_\uparrow$ and $N_\downarrow$. Adapted from Zwierlein *et al.* [171].

finite temperature Thomas-Fermi profile. In contrast, below the critical trap depth (lower panels), the shape of the smaller cloud start to deviate drastically from the Thomas-Fermi distribution of a normal gas.

By using the wings of the larger component, which are non-interacting, as a thermometer, Zwierlein *et al.* [171] determined the critical temperature for the phase transition. They reported $T/T_F = 0.18$ for the BEC side at imbalance $p = 75 \pm 3\%$, $T/T_F = 0.12$ for the resonance at $p = 59 \pm 3\%$, and $T/T_F = 0.11$ for the BCS side at $p = 53 \pm 3\%$. Notably the critical temperature will in general depend on the population imbalance. For example, for large enough imbalance on resonance or on the BCS side, no condensate will form even at zero temperature and the critical temperature for superfluidity will be zero.

Shin *et al.*[172] observed phase separation between the superfluid and the normal component in a strongly interacting Fermi gas with imbalanced spin populations. The density difference between the two spin components is directly measured *in situ* using a special phase-contrast imaging technique as shown in Figure 4.8. For a partially superfluid im-

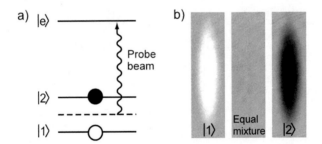

Figure 4.8. Phase-contrast imaging of the density difference of two spin states. (a) The probe beam is tuned to the red for the $|1\rangle \rightarrow |e\rangle$ transition and to the blue for the $|2\rangle \rightarrow |e\rangle$ transition. The resulting optical signal in the phase-contrast image is proportional to the density difference $n_d \equiv n_1 - n_2$, where n_1 and n_2 are the densities of the states $|1\rangle$ and $|2\rangle$, respectively. (b) Phase-contrast images of trapped atomic clouds in state $|1\rangle$ (left) and state $|2\rangle$ (right), and of an equal mixture of the two states (middle). Adapted from Shin *et al.* [172].

balanced mixture, a shell structure was observed in the *in situ* phase-contrast images (Figure 4.9, top panel). Since the image shows the column density difference (the 3D density difference integrated along the y-direction of the imaging beam), the observed depletion in the center indicates a 3D shell structure with even stronger depletion in the central region. The size of this inner core decreases for increasing imbalance and the core shows a distinctive boundary until it disappears for large imbalance (see central panel of Figure 4.9).

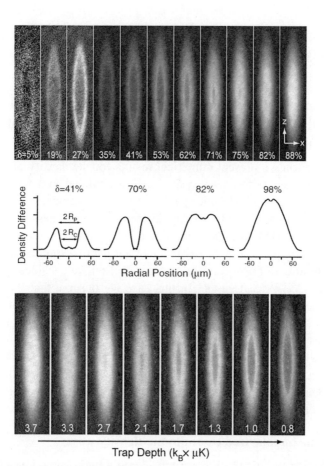

Figure 4.9. Top panel: *In situ* direct imaging of trapped Fermi gases with population imbalance. For $p \leq 75\%$, a distinctive core was observed showing the shell structure of the cloud. Central panel: Characterization of the shell structure by reconstructed 3D radial profiles at B=834 G. R_p and R_c show the peak position of the magnetization and radius of the empty core. Bottom panel: Emergence of phase separation in an imbalanced Fermi gas. The temperature of the cloud was controlled by varying the final value of the trap depth in the evaporation process. Adapted from Shin *et al.* [172].

In situ phase-contrast images of a cloud at various temperatures are shown in bottom panel of Figure 4.9. The temperature T of the cloud is controlled with final value of the trap depth in the evaporation process. The shell structure appears and become prominent when T decreases below a certain critical value. This shell structure gives rise to a bimodal density profile of the minority component that was observed after expansion from the trap in the previous experiment by Zwierlein *et al.* [171]. Here

Shin *et al.* [172] showed by *in situ* measurements that the onset of superfluidity is accompanied by a pronounced change in the spatial density
profile.

4.4. Numerical results for attractive fermions trapped in 1D optical lattices from the Bethe-Ansatz-based local-density approximation

In this section we present numerical results for the density profiles of a
1D Fermi gases with attractive interaction confined in a harmonic potential inside an optical lattice. In the first part we review the work by
Gao Xianlong *et al.* [150] on unpolarized systems. These authors have
proposed that 1D Fermi gases in optical lattices can be used to study the
Luther-Emery phase and the existence of antiparallel spin pairing Luther-
Emery liquid. In the second part of this section we consider spin polarized (*i.e.* $N_\uparrow \neq N_\downarrow$) gases. In this case we observe that the majority
atoms (say spin-up) move to the edges of the trap while a core contain-
ing both species is formed in the bulk of the trap. The size of the core
decreases with increasing polarization p.

We will present also numerical results obtained by means of the DMRG
technique. These results which have been obtained using the DMRG
code released within the "Powder with Power" project,[1] have been kindly
provided to us by Matteo Rizzi and Prof. R. Fazio.

4.4.1. The unpolarized case: atomic-density waves

Gao Xianlong *et al.* [150] have considered a two-component Fermi gas
with N atoms confined by a harmonic potential of strength V_2 in a $1D$
lattice with unit lattice constant and L lattice sites. Similarly to the model
considered in Chapter 3 (see Section 3.6.1), the system is described in
Ref. [150] by an inhomogeneous Fermi-Hubbard Hamiltonian,

$$\hat{\mathcal{H}} = -t \sum_{i,\sigma} \left[\hat{c}_\sigma^\dagger(z_i)\, \hat{c}_\sigma(z_{i+1}) + \text{H.c.} \right]$$
$$+ U \sum_i \hat{n}_\uparrow(z_i)\hat{n}_\downarrow(z_i) + V_2 \sum_{i,\sigma} (z_i - L/2)^2\, \hat{n}_\sigma(z_i). \quad (4.14)$$

This Hamiltonian with repulsive interaction ($U > 0$) has been discussed
in Chapter 3. Here we focus on attractive interaction ($U < 0$). Within the
BALDA scheme presented in Chapter 3, and in parallel with a DMRG
study Gao Xianlong *et al.* have calculated the ground-state properties of

---

[1] Consult the web site http://www.qti.sns.it

the Hamiltonian (4.14). These authors showed the tendency to pairing by analyzing the pair binding energy defined as

$$E_P = E_{GS}(N + 2) + E_{GS}(N) - 2\, E_{GS}(N + 1) \qquad (4.15)$$

Table 4.1 reports results for $N = 30$ fermions in a lattice with $L = 100$ sites, inside a trap with $V_2/t = 4 \times 10^{-3}$. E_{GS} is negative, signaling a tendency to pairing.

Table 4.1. Ground-state and pair-binding energies for $N = 30$, $L = 100$, and $V_2/t = 4 \times 10^{-3}$. The agreement between DMRG and BALDA for E_{GS} is quite satisfactory even for $u = -20$, where the deviation is about 2.2%. However, BALDA tends to overestimate E_P even at moderate values of u. The "$\times$" sign indicates that the spin-BALDA code for 31 atoms does not converge for $u = -20$. From Gao Xianlong et al. [150].

u	$E_{GS}^{BALDA}/(tL)$	$E_{GS}^{DMRG}/(tL)$	E_P^{BALDA}/t	E_P^{DMRG}/t
-0.5	-0.35824	-0.35832	-0.0283	-0.0213
-1	-0.39336	-0.39340	-0.0614	-0.0577
-2	-0.47672	-0.47631	-0.3265	-0.2442
-4	-0.70693	-0.69010	-5.1008	-1.3278
-20	-3.05320	-2.98536	$\times$	-16.4217

The pairing is associated with the presence of Atomic-Density Waves (ADWs), which are stabilized by the harmonic potential. Figure 4.10 reports the numerical results for the site occupation of a gas with $N = 30$ atoms. The consequences of Luther-Emery pairing in the presence of confinement are dramatic. For $u < 0$ the site occupation exhibits a density wave (with $N/2$ peaks in a weak trap), reflecting the tendency of atoms with different pseudospins to form stable spin-singlet dimers that are delocalized over the lattice.

For small V_2/t (see the top panel of Figure 4.10) $n(z_i)$ in the bulk of the trap ($80 \leq i \leq 100$) can be fitted to an ADW of the form $n_i = \tilde{n} + A_{ADW}\cos(k_{ADW}\, i + \varphi)$. For example, for $V_2/t = 10^{-5}$ the authors have found $k_{ADW} = 0.73$ for $u = -1$ and $k_{ADW} = 0.84$ for $u = -3$. In such a weak confinement the oscillations of the site occupation extend to regions far away from the center of the trap, where they are characterized by smaller edge wavenumbers. For $V_2 = 0$ bosonization predicts [44] an

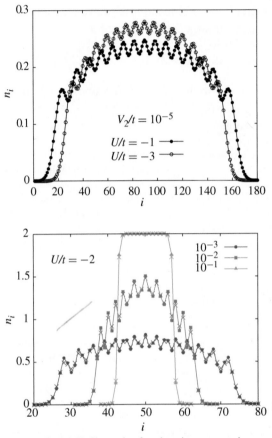

Figure 4.10. Top panel: DMRG results for the site occupation n_i as a function of site position i for a system with $N = 30$ fermions in $L = 180$ lattice sites, and in the presence of a harmonic potential with $V_2/t = 10^{-5}$. For this value of V_2/t BALDA overestimates the ADW amplitude. Bottom panel: DMRG results (crosses) for $N = 30$, $L = 100$, and $u = -2$ are compared with BALDA data (filled symbols). V_2/t is increased from 10^{-3} to 10^{-1}. The thin solid lines are just a guide for the eye. Adapted from Gao Xianlong [150].

incipient ADW with wavenumber $2k_F$ quenched by strong quantum fluctuations. In the present case they found $k_{ADW} \simeq 2k_F^{eff}$, where the effective Fermi wavenumber $k_F^{eff} = \pi \widetilde{n}/2$ is determined by the average density in the bulk of the trap (note that $k_{ADW} = \pi$ when the average density in the bulk reaches half filling).

Finite-size effects become important on increasing V_2 (see the bottom panel of Figure 4.10) and a simple fitting formula such as the one used above does not work even at the center of the trap. Eventually when

$V_2/t \approx 10^{-1}$ a region of doubly-occupied sites develops at the center of the trap: spin-singlet dimers, which in a weak trap are delocalized, are squeezed close together to produce an extended region of $\approx N/2$ doubly-occupied consecutive sites.

Figure 4.11 shows how the ADWs evolve with increasing $|u|$ at fixed V_2/t. For weak-to-intermediate coupling ADWs are present in the bulk of the trap. The agreement between the BALDA and the DMRG results is excellent for $|u| \leq 1$. With increasing $|u|$ the BALDA scheme deteriorates. Since the $U < 0$ Bethe *Ansatz* is obtained from that at $u > 0$ by means of an exact transformation, this shows that the locality assumption inherent in the LDA is less well satisfied for attractive than for repulsive interactions. In fact, it appears that the BALDA performance for a fixed value of $|u|$ depends on the value of V_2/t. This leads to an overestimation of the amplitude of the ADWs (see panel B). According to DMRG, the bulk ADWs disappear in the extreme strong-coupling limit (see panel C and D). For $|u| \gg 1$ a flat region of doubly-occupied sites emerges at the trap center, resembling that described above for the case of weak interactions and strong confinement (see the bottom panel of Figure 4.10).

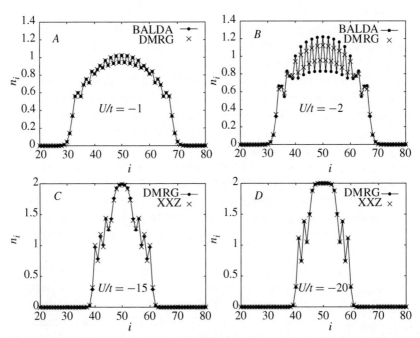

Figure 4.11. Site occupation n_i as a function of i for $N = 30$, $L = 100$, and $V_2/t = 4 \times 10^{-3}$. Panels A and B: DMRG results (crosses) are compared with BALDA data (filled circles). Panels C and D: DMRG results for the Hamiltonian (4.14) (filled circles) are compared with DMRG results for the strong-coupling Hamiltonian (4.16) (crosses). The thin solid lines are just a guide for the eye. Adapted from Gao Xianlong *et al.* [150].

The disappearance of the ADWs at strong coupling can be explained by mapping the Hamiltonian (4.14) onto a spin-1/2 XXZ model [44, 51],

$$\hat{\mathcal{H}}_\infty = \sum_i \sum_{\alpha=x,y,z} J_\alpha \hat{\sigma}_i^\alpha \hat{\sigma}_{i+1}^\alpha + \sum_i B_i \hat{\sigma}_i^z, \qquad (4.16)$$

with $J_x = J_y = -J_z = -t^2/|U|$ and $B_i = V_2(i - L/2)^2$. The total site occupation operator $\hat{n}_i$ is related to $\hat{\sigma}_i^z$ by $\hat{n}_i = 1 + \hat{\sigma}_i^z$. Particle-number conservation requires working in a sector with fixed total magnetization $\langle \sum_i \hat{\sigma}_i^z \rangle = N - L \equiv M$. In the limit $|u| \to \infty$, J_α is negligibly small and thus finding the ground state of $\hat{\mathcal{H}}_\infty$ is equivalent to solving the problem of orienting a collection of spins in a nonuniform magnetic field in order to minimize the Zeeman energy in producing a magnetization M. Thus, for $|U|/t \to \infty$ one expects a classical state with $\langle \hat{\sigma}_i^z \rangle = 1$ ($\langle \hat{n}_i \rangle = 2$) in $N/2$ sites at the trap center where B_i is small, and $\langle \hat{\sigma}_i^z \rangle = -1$ ($\langle \hat{n}_i \rangle = 0$) in the remaining $L - N/2$ sites.

4.4.2. Phase separation in spin-polarized gases

In this section we consider an attractive polarized Fermi gas confined by a harmonic potential inside a 1D optical lattice. We employ the BALSDA scheme to obtain ground-state spin-resolved density profiles. We solve equations (3.68)-(3.72) for attractive interactions ($u < 0$) in order to calculate $\varepsilon_{gs}(n, s; u)$. Therefore we can build $V_{xc}^{hom}(n, s; u)$ and based on BALSDA, equation (3.91), we calculate the exchange-correlation potential of the inhomogeneous Hubbard model. This potential completes the approach for solving Kohn-Sham equations (3.87)-(3.89).

In Figures 4.12-4.14 we show the ground-state occupations in a spin polarized Fermi gas with attractive interactions $u = -1$ to $u = -3$ in a lattice with $L = 100$ sites. In all cases the BALSDA results show very good agreement with the DMRG data. For each u spin-up and spin-down the density profiles oscillate in phase (contrary to $u > 0$, see Section 3.6.2), showing $N_\uparrow$ and $N_\downarrow$ peaks respectively. Furthermore the atoms are more confined in the center of the trap for stronger value of the interaction ($u = -3$).

As we have seen in Section 4.4.1, in the unpolarized gas for N atoms there are $N/2$ pairs, reflecting in $N/2$ peaks in the ADWs. In the present polarized situation some atoms of majority species remain without pairing atoms. We observe this fact in Figures 4.12-4.14: in all cases there is a central part of quasi-flat magnetization in the portion of the lattice where spin-down atoms are present. On the other hand spin-up atoms accumulate at the edges of the trap. There are $N_\downarrow$ delocalized dimers which give rise to $N_\downarrow$ peaks in the total density profiles. The attractive

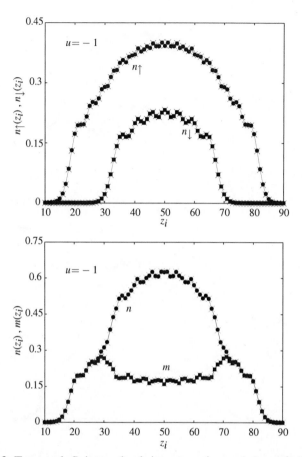

Figure 4.12. Top panel: Spin-resolved site occupation $n_\sigma(z_i)$ as a function of z_i by DMRG (crosses) and BALSDA (filled symbols) for a system with attractive interaction $u = -1$, $N_\uparrow = 20$ spin-up atoms and $N_\downarrow = 7$ spin-down atoms in $L = 100$ lattice sites, and in the presence of a harmonic potential with $V_2/t = 0.001$. Bottom panel: Total density profiles $n(z_i)$ and local magnetization $m(z_i)$ as a function of z_i for the same system. The thin solid lines are just a guide for the eye.

interaction enhances the formation of these dimers and in Figure 4.14 we see this fact in the higher value of total density n and, more significantly, by a clear appearance of the ADW in the total density distribution.

In Figure 4.15 we show a 3D plot of magnetization for different values of polarization p. We observe that for each value of polarization there is a core of coexisting atoms, however this becomes smaller and smaller for higher value of p and finally disappears in the completely polarized gas at $p = 1$. We notice that in all cases there is a *finite* value for the

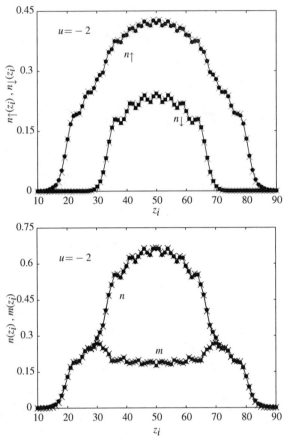

Figure 4.13. Top panel: Spin-resolved site occupation $n_\sigma(z_i)$ as a function of z_i by DMRG (crosses) and BALSDA (filled symbols) for a system with attractive interaction $u = -2$, $N_\uparrow = 20$ spin-up atoms and $N_\downarrow = 7$ spin-down atoms in $L = 100$ lattice sites, and in the presence of a harmonic potential with $V_2/t = 0.001$. Bottom panel: Total density profiles $n(z_i)$ and local magnetization $m(z_i)$ as a function of z_i for the same system. The thin solid lines are just a guide for the eye.

magnetization in the bulk of the trap. This value indicates that some unpaired spin-up atoms can cross the core where both species coexist. Concerning the experiment, we should emphasize that the central panel of Figure 4.9 shows the experimental evidence for phase separation in the 3D Fermi gas at the strongly interacting regime. This regime leads to the quasi-equal populations for both components in the bulk of the trap, *i.e.* a near-zero central region in density difference. For the results in Figure 4.12-4.14 the attractive interaction is finite so there is no near-zero magnetization in the bulk of the trap.

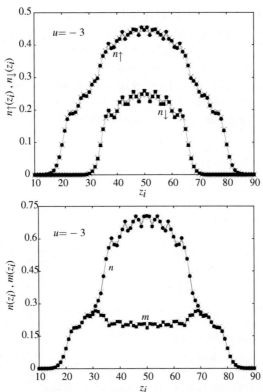

Figure 4.14. Top panel: Spin-resolved site occupation $n_\sigma(z_i)$ as a function of z_i by DMRG (crosses) and BALSDA (filled symbols) for a system with attractive interaction $u = -3$, $N_\uparrow = 20$ spin-up atoms and $N_\downarrow = 7$ spin-down atoms in $L = 100$ lattice sites, and in the presence of a harmonic potential with $V_2/t = 0.001$. Bottom panel: total density profiles $n(z_i)$ and local magnetization $m(z_i)$ as a function of z_i for the same system. The thin solid lines are just a guide for the eye.

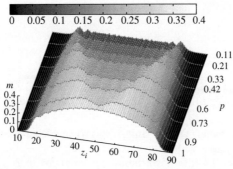

Figure 4.15. 3D plot for magnetization m as a function of z_i and for different values of p. The attractive interaction is $u = -2$, $L = 100$ and the harmonic potential has strength $V_2/t = 0.001$.

In Figure 4.16 we show DMRG results for the spin-resolved density profiles at strong coupling $u = -200$. We observe that in this case both
density profiles are nearly-equal in the center of the trap (top panel of
Figure 4.16). This equality results in near-zero magnetization in the center (bottom panel of Figure 4.16). Figure 4.17 shows the local magnetization for different values of the polarization p. We see that in the bulk of
the trap there is a finite value of magnetization which increases for higher
values of p.

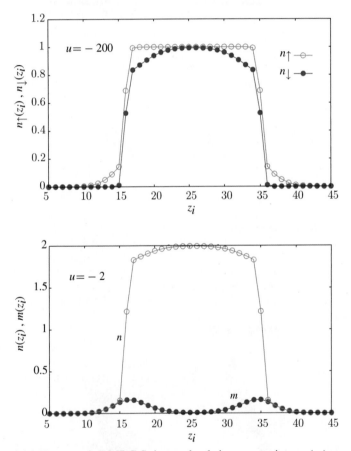

Figure 4.16. Top panel: DMRG Spin-resolved site occupation $n_\sigma(z_i)$ as a function of z_i for a system with attractive interaction $u = -200$, $N_\uparrow = 20$ spin-up
atoms and $N_\downarrow = 18$ spin-down atoms ($p = 0.05$) in $L = 50$ lattice sites,
and in the presence of a harmonic potential $V_{ext}(z_i) = V_2(z_i - \frac{L+1}{2})^2$ with
$V_2/t = 2.25 \times 10^{-3}$. Bottom panel: total density profiles $n(z_i)$ and local magnetization $m(z_i)$ as a function of z_i for the same system. The thin solid lines are
just a guide for the eye.

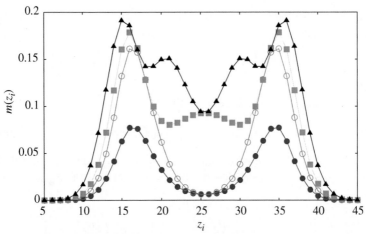

Figure 4.17. DMRG results for the local magnetization $m(z_i)$ as a function of z_i for $u = -200$ in $L = 50$ lattice sites, and in the presence of a harmonic potential $V_{ext}(z_i) = V_2 (z_i - \frac{L+1}{2})^2$ with $V_2/t = 2.25 \times 10^{-3}$. The population of spin-up atoms is $N_\uparrow = 20$ while spin-down population varies to $N_\downarrow = 19$ (blue), $N_\downarrow = 18$ (red), $N_\downarrow = 17$ (green) and $N_\downarrow = 16$ (black). These populations respectively correspond to polarization $p = 0.02, 0.05, 0.08$ and 0.11. The thin solid lines are just a guide for the eye.

4.5. The Bogoliubov-de Gennes approximation for attractive fermions trapped in 1D optical lattices

In this section we analyze 1D attractive Fermi gases in a lattice within a mean-field approximation. This formalism leads to discrete Bogoloiubov-de Gennes (BdG) equations. This mean-field approach, however, ignores exchange-correlation effects which have been treated in Chapter 3 via a suitable local-density approximation. A mixed scheme incorporating both off-diagonal BdG and xc effects, such as in the DFT for superconductors [178], would be highly desirable. The BdG equations provide spin-resolved density profiles and the off-diagonal pairing gap.

4.5.1. Bogoliubov-de Gennes formalism in 1D lattices.

To describe 1D Fermi gases confined by an external harmonic potential inside a optical lattice, we rewrite the Hubbard model in the form

$$
\begin{aligned}
\hat{\mathcal{H}} = -t \sum_{i,\sigma} & \left[\hat{c}_\sigma^\dagger(z_i)\, \hat{c}_\sigma(z_{i+1}) + \hat{c}_\sigma^\dagger(z_{i+1})\, \hat{c}_\sigma(z_i) \right] \\
& + U \sum_i \hat{n}_\uparrow(z_i)\hat{n}_\downarrow(z_i) + \sum_i \left[V_{ext}(z_i) - \mu_\sigma \right] \hat{n}(z_i).
\end{aligned}
\tag{4.17}
$$

In this Hamiltonian the interaction term in the mean-field approximation can be written as

$$\hat{\mathcal{H}}_I = \sum_i \left[\Delta_i \, \hat{c}_\uparrow^\dagger(z_i) \, \hat{c}_\downarrow^\dagger(z_i) - \Delta_i^* \, \hat{c}_\uparrow(z_i) \, \hat{c}_\downarrow(z_i) \right], \qquad (4.18)$$

where the pairing gap Δ is defined by

$$\Delta_i \equiv \Delta(z_i) = -U \, \langle \, \hat{c}_\sigma(z_i) \, \hat{c}_{\bar{\sigma}}(z_i) \, \rangle. \qquad (4.19)$$

Performing the BdG transformation [179] and requiring the Hamiltonian to be diagonal, one can obtain the following lattice BdG equations [180, 181]:

$$\sum_j \begin{pmatrix} \mathcal{H}_{ij,\sigma} & \Delta_{ij}^* \\ \Delta_{ij} & -\mathcal{H}_{ij,\bar{\sigma}}^* \end{pmatrix} \begin{pmatrix} u_\sigma^n(z_j) \\ v_{\bar{\sigma}}^n(z_j) \end{pmatrix} = E_n \begin{pmatrix} u_\sigma^n(z_j) \\ v_{\bar{\sigma}}^n(z_j) \end{pmatrix}, \qquad (4.20)$$

where the single-particle Hamiltonian $\mathcal{H}_{ij,\sigma} = -t\delta_{i,i\pm1} - \left(V_{\text{ext}}(z_i) - \mu_\sigma\right)\delta_{i,j}$ and $\Delta_{ij} = \Delta_i \, \delta_{ij}$. Here $u_\sigma^n(z_j)$, $v_{\bar{\sigma}}^n(z_j)$ are the Bogoliubov quasi-particle amplitudes on the j-th site. We solve self-consistently BdG equations (4.20). The self-consistent conditions for the number of atoms and the pairing gap are [182]

$$n_\uparrow(z_i) = \sum_{n=1}^{2L} |\mathbf{u}^n(z_i)|^2 \, f(E_n) \qquad (4.21)$$

$$n_\downarrow(z_i) = \sum_{n=1}^{2L} |\mathbf{v}^n(z_i)|^2 \, [1 - 2f(E_n)] \qquad (4.22)$$

$$\Delta_{ij} = \frac{U}{4} \sum_{n=1}^{2L} \left(\mathbf{u}^n(z_i) \, \mathbf{v}^{*n}(z_j) + \mathbf{v}^{*n}(z_i) \, \mathbf{u}^n(z_j) \right) \tanh(E_n/2k_{\mathrm{B}}T) \quad (4.23)$$

where $\mathbf{u}^n(z_i) = (-v_\uparrow^{*n}(z_i), u_\uparrow^n(z_i))$ and $\mathbf{v}^n(z_i) = (u_\uparrow^{*n}(z_i), v_\downarrow^n(z_i))$ are the row vectors, and $f(E)$ is the Fermi-Dirac distribution function and the atom-number normalization reads as $N_\sigma = \sum_i n_\sigma(z_i)$. In the following section we analyze the results for the densities and gap obtained by these equations.

4.5.2. Behavior of the pairing gap and LOFF phase in the spin-polarized case

In Figures 4.18-4.20 we show the results for spin-resolved density profiles and pairing gap for a Fermi gas inside a 1D lattice and confined by an harmonic external potential $V_{\text{ext}}(z_i) = V_2(z_i - L/2)^2$. The temperature is fixed at $T = 10^{-3} t/k_{\mathrm{B}}$. We have solved self-consistently the

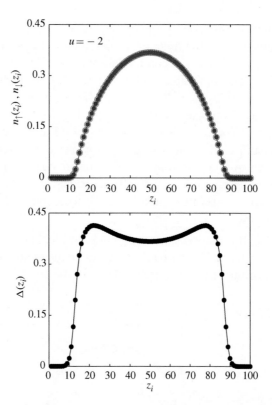

Figure 4.18. Top panel: spin-up (red) and spin-down (blue) densities for an unpolarized Fermi gas with $N_\uparrow = 20$, $N_\downarrow = 20$ in a lattice site with $L = 100$ sites and under an harmonic confinement with $V_2/t = 0.001$. Bottom panel: local pairing gap $\Delta(z_i)$ in units of t as a function of lattice coordinates z_i for the same system.

discrete BdG equations (4.21)-(4.23) which give the spin-resolved $n_\sigma(z_i)$ and $\Delta(z_i)$.

Even though strictly speaking a mean-field gap Δ is not well-defined in a 1D system (see Section 4.2) the BdG results for the polarized dependence of Δ are very suggestive and are shown in Figures 4.18-4.20. Figure 4.18 presents the results for $n_\sigma(z_i)$ (top panel) and $\Delta(z_i)$ (bottom panel) in a system consisting of $N_\uparrow = N_\downarrow = 20$ and $u = -2$. Because of zero-polarization it is possible for all atoms to be paired. This fact is reflected in the behavior of position-dependent pairing gap $\Delta(z_i)$. Due to the inhomogeneous density distribution in the trapped gas a spatially varying pairing gap is expected [183, 184]. In the bulk of the trap the pairing gap keeps a definite sign. Here we call it the "BCS" state which has a definite sign in the gap function [184]. The gap function decays and goes to zero exponentially near the edges of the trap.

The behavior of the pairing gap $\Delta(z_i)$ and spin-resolved densities change significantly in the finite polarization case. Physically for $p \neq 0$ there is no room to neatly accommodate the excess majority species in the BCS state. The sign of $\Delta(z_i)$ must change to accommodate the excess majority species. We call this the "LOFF" region where, unlike the BCS state, the pairing gap changes its sign. Figure 4.19 presents results for a system that contains $N_\uparrow = 20$ and $N_\downarrow = 15$ atoms, corresponds to relatively small polarization $p = 0.14$. It shows the presence of a BCS core where the spin-up and spin-down atoms are nearly equal in number and balanced. The gap develops fully there and near the edges of the trap ($z_i > 70$) changes its sign, allowing it to accommodate the excess majority species.

In $D = 3$ LOFF pairing starts only after a critical polarization. In our results the oscillations in the pairing gap and density are also visible for small polarization. This is understandable in the sense that, as the trap

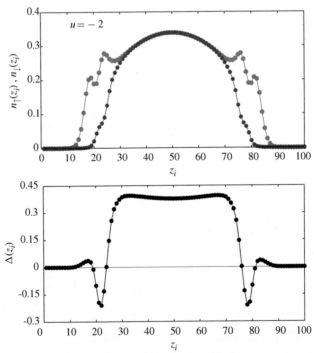

Figure 4.19. Top panel: spin-up (red) and spin-down (blue) densities for an imbalanced Fermi gas with $N_\uparrow = 20$, $N_\downarrow = 15$ atoms ($p = 0.14$) in a lattice site with $L = 100$ sites and under an harmonic confinement with $V_2/t = 0.001$. Bottom panel: local pairing gap $\Delta(z_i)$ in units of t as a function of lattice coordinates z_i for the same system.

favors phase separation, the *local* polarization $p(z_i) = n_\uparrow(z_i) - n_\downarrow(z_i)$ at the edges of the trap becomes very easily of considerable size even when the *global* polarization p is small. Therefore, locally one can fulfill the LOFF condition of exceeding a critical polarization [185]. One could interpret the results in the following way: the trap tends to enforce a normal BCS state at the center of the trap and a LOFF state at the edges and the significance of the latter grows with the total polarization.

Figure 4.20 shows the results for the spin-resolved density and the pairing gap for $N_\uparrow = 20$ and $N_\downarrow = 3$ atoms, corresponds to the relatively high polarization $p = 0.73$. The oscillations in the pairing gap are more evident and we observe them in the pairing region, *i.e.* where $n_\downarrow(z_i) \neq 0$. In the gap oscillations the spin-down distribution plays a more important

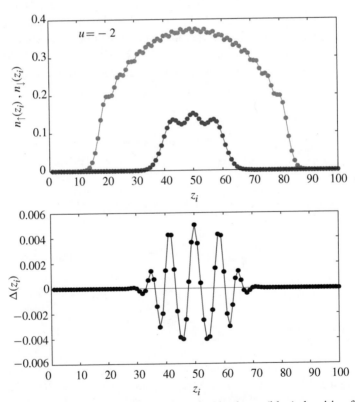

Figure 4.20. Top panel: spin-up (red) and spin-down (blue) densities for an imbalanced Fermi gas with $N_\uparrow = 20$, $N_\downarrow = 3$ atoms ($p = 0.73$) in a lattice site with $L = 100$ sites and under an harmonic confinement with $V_2/t = 0.001$. Bottom panel: local pairing gap $\Delta(z_i)$ in units of t as a function of lattice coordinates z_i for the same system.

role. The reason being that the majority spin-up atoms are spread on the lattice therefore the minority spin-down atoms, which are mainly in the bulk of the trap, determine the extrema of the pairing gap. At the center of the trap ($z_i = 50$), the spin-down density has a maximum value and spin-up has the minimum value, therefore there is strong pairing between spin-up and spin-down atoms which leads to a peak in $\Delta(z_i)$. By moving from the center of the trap to the right, the spin-down density decreases and reaches a minimum value at $z_i = 54$ while the spin-up reaches a maximum value. Therefore because of lack of minority spin-down species there is less chance of pairing and so we observe a minimum in $\Delta(z_i)$ at this position. The oscillating behavior of the gap can be explained by this qualitative picture.

Finally in Figure 4.21 we show the pairing gap $\Delta(z_i)$ for different values of the polarization. It can be seen that starting from unpolarized gas, the gap $\Delta(z_i)$ evolves from a BCS-like state to the LOFF state. The modulation of the pairing gap and the presence of the nodes are signals which indicate that the imbalanced Fermi gas is in a LOFF phase [176]. The effect of the polarization is mainly to decrease the superfluid core size and also to decrease the amplitude of the gap oscillations.

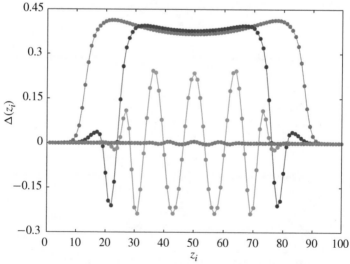

Figure 4.21. Local pairing gap $\Delta(z_i)$ in units of t as a function of lattice coordinates z_i for different values of polarization $p = 0$ (red), $p = 0.14$ (blue), $p = 0.42$ (green) and $p = 0.73$ (magenta).

Conclusions

In summary, we reviewed in Chapter 1 some fundamental aspects of the physics of atomic quantum gases. The Chapter is meant to provide all the background which is needed to understand the original developments reported in the following Chapters.

In Chapter 2 we have focused attention on some main features of the coherent transport of matter by quantum gases through linear arrays of potential wells under a constant external drive. Transport by Fermi-surface electrons, by a Bose-Einstein-condensed atomic gas, and by a spin-polarized atomic Fermi gas have been comparatively discussed for arrays of identical or alternating paired wells and for a Fibonacci-ordered sequence, with the main aim of exposing the emergence of localization from band-structure effects and from quasi-periodic disorder. The decimation/renormalization procedure that we have briefly recalled is, of course, of more general usefulness. In particular, we refer at this point to a very recent study of electronic transport and the emergence of a metal-insulator transition in a chaotic 1D sequence of energy levels by Pinto *et al.* [186].

In comparison with experiments on electronic transport through arrays or networks of confining sites such as quantum dots, the experimental study of transport by quantum atomic or molecular gases through optical arrays is still in its initial stages. In regard to Bloch oscillation experiments [96, 101], bosons or fermions go through a single-well array (and leave it in the form of regular drops of matter when access to vacuum is open) with a precise periodicity related to the driving force by h/Fd. The emergence of drops is an unambiguous manifestation of long-range coherence, and the number of particles in each drop increase monotonically with the strength of the drive [106]. Transport of atomic quantum matter through other types of array remains to be studied experimentally. In the case of a double-well array, the relation of the characteristic oscillation period to the drive becomes $h/(2Fd)$ and a second characteristic time arises from inter-sub-band tunneling. Transport by matter waves is in this situation entirely equivalent to interference by beam splitting in optics, as we record in Appendix A.

Finally, matter wave transport through a quasi-periodic chain is governed by the fragmentation of the energy spectrum due to disorder. This and other important aspects of transport theory could be profitably studied in experiments on atomic or molecular gases moving inside optical arrays.

In Chapter 3 we investigated the ground state of two-component Fermi gases with repulsive interactions subject to external confining potentials inside 1D optical lattices. We have used a novel version of spin-density-functional theory in which a local-spin-density approximation based on the Bethe-Ansatz solution for the 1D Hubbard model is used. We have obtained results for spin-resolved atom-density profiles in two different situation: (i) a spin-polarized system ($N_\uparrow \neq N_\downarrow$) in a single harmonic trap, and (ii) a spin-polarized system in spin-dependent harmonic traps ($V_2^\uparrow \neq V_2^\downarrow$). In case (i) the results show spin density waves in the center of the trap. The repulsive intercomponent interaction increases the amplitude of the spin-density waves and also broadens the spin-resolved density profiles. Our results are in very good agreement with DMRG data, which are believed to be essentially exact, in a weak-to-intermediate interaction regime. In case (ii) for sufficiently large difference in the strength of the traps and in an intermediate interacting regime, we have observed that the weakly confined minority-spin atoms move totally to the edges of the trap. This complete "phase-separation" between two components is driven by the on-site repulsive interaction.

In Chapter 4 we have studied Fermi gases with attractive interactions subject to parabolic trapping inside 1D optical lattices. With the same approach of spin-density-functional theory used in Chapter 3, we have found the spin-resolved atomic density profiles for spin imbalanced Fermi gases. In the intermediate interacting regime ($u \approx -3$), our results show an accumulation of unpaired majority-spin atoms at the edges of the trap while the paired atoms form a core at the center of the trap. In respect to the density profiles of the unpolarized gas, the results for spin-resolved density profiles show better agreement with DMRG data. We could conclude that the local-spin-density approximation is more reliable in the polarized system. We have studied the polarized Fermi gas also in the mean-field level by solving the Bogoliubov-de Gennes equations. These equations determine the density profiles and also the local pairing gap. The pairing gap in the unpolarized situation shows BCS-like behavior and remains flat while even for a small amount of spin polarization it changes sign and starts to oscillate. The oscillations appear first at the edges of the trap and with increasing spin polarization they show up everywhere in the trap. The pairing gap eventually approaches zero in the fully-spin-polarized gas. This behavior is expected by the general argument of a position-dependent pairing gap in the LOFF phase.

Appendix A
An example of matter-wave optics

We showed in Section 2.4.4 the propagation of condensate wavepacket in a double-well lattice. In this appendix, we mention its optical counterpart in which a light beam travells through a five-layer optical medium.

A condensate wavepacket propagating in a double-well lattice is equivalent to a light beam travelling through a five-layer optical medium [110]. The first and last medium in Figure A.1 are semi-infinite and play the role

Figure A.1. Birefringent system for light propagation, representing a two-band array connected to an incoming and an outgoing lead.

of the leads. The second and fourth layer stand for the two energy sub-bands, while the middle layer mimics the inter-band minigap.

Let $\theta_{i,j}$ and $r_{i,j}$ be the transmission and reflection coefficient at the interface between media i and j, with $\theta_{i,j} = 1 + r_{i,j}$. For a suitable coupling between the leads and the lattice, $\theta_{1,2}$ and $\theta_{4,5}$ have unit modulus and the wave takes up an irrelevant phase factor on crossing these interfaces. At the 2-3 and 3-4 interfaces we can mimic the effect of the minigap in Bragg-reflecting the condensate by imposing total reflection of light *via* $r_{2,3} = \exp(i\alpha_{2,3})$ and $r_{3,4} = \exp(i\alpha_{3,4})$, and allow for tunnel through the minigap by propagation through layer 3 *via* an evanescent wave. If media 2 and 4 have the same refractive index and the same optical depth, we have $\alpha_{3,4} = \alpha_{2,3} - \pi$ and by symmetry we can set $\alpha_{2,3} = -\alpha_{3,4} = \pi/2$. We can then use the recursive formula [187]

$$\begin{cases} r_{i,j+2} = \dfrac{r_{i,j+1} + r_{j+1,j+2}e^{2i\alpha_{j+1}}}{1 + r_{i,j+1}r_{j+1,j+2}e^{2i\alpha_{j+1}}} \\[2ex] \theta_{i,j+2} = \dfrac{\theta_{i,j+1} + \theta_{j+1,j+2}e^{i\alpha_{j+1}}}{1 + r_{i,j+1}r_{j+1,j+2}e^{2i\alpha_{j+1}}} \end{cases} \tag{A.1}$$

to calculate the transmission coefficient between media 1 and 5, where α_j is the phase shift acquired by light travelling through the j-th layer. To pursue the analogy with the double-well lattice we set $\alpha_2 = \alpha_4 = \omega T_B/2$ and $\alpha_3 = i\omega\tau_t/2$, with $\tau_t = 2\pi/\omega$ and ω being the light-beam frequency. The total transmission coefficient for the light intensity is $|\theta_{1,5}|^2$ and is found to have minima when the ratio T_B/τ_t takes integer values, as in the case of matter waves in a double-well lattice.

The main difference between the matter-wave and light-wave patterns is that in the case of a Bose-Einstein condensate the height of the transmission peaks is largest at low values of T_B/τ_t (see Figure 2.14), whereas in the optical analog all peaks have the same height. Decreasing T_B leads to a decrease of the optical depth of media 2 and 4 in the five-layer system, while for a condensate it means that the bosons leave the lattice towards the continuum after having travelled through a lower number of sites. Decreasing T_B is equivalent in this case to shortening the array and hence helps the tunnelling.

Appendix B
The internal magnetic field in the 1D homogeneous Hubbard model from the Bethe-Ansatz solution

The internal magnetic field for the homogeneous Hubbard model is defiend as

$$h(n, s; u) = \frac{\partial}{\partial s}\varepsilon_{GS}(n, s; u), \tag{B.1}$$

and by equation (3.72) for the ground state energy we have

$$h(n, s; u) = -u\Theta(-u) - 2t \int_{-Q}^{Q} dk\rho_s(k)\cos(k) - 4t\, Q_s\, \rho(Q)\cos Q. \tag{B.2}$$

Here $\Theta(x) = 1$ if $x > 0$ and $\Theta(x) = 0$ if $x < 0$; $\rho_s(k) \equiv \partial\rho(k)/\partial s$ and $\sigma_s(\lambda) \equiv \partial\sigma(\lambda)/\partial s$ satisfy the following integral equations [188, 189]:

$$\rho_s(k) = \frac{u\cos k}{4\pi} \int_{-B}^{B} d\lambda\sigma_s(\lambda)f_1(k, \lambda) \tag{B.3}$$

$$+ \frac{u\cos k}{4\pi}[f_1(k, B) + f_1(k, -B)]\sigma(B)B_s,$$

$$\sigma_s(\lambda) = \frac{u\cos k}{4\pi} \int_{-Q}^{Q} dk\rho_s(k)f_1(k, \lambda) \tag{B.4}$$

$$- \frac{u\cos k}{2\pi} \int_{-B}^{B} d\lambda'\sigma_s(\lambda')f_2(\lambda, \lambda')$$

$$+ \frac{u\cos k}{4\pi}[f_1(Q, \lambda) + f_1(-Q, \lambda)]\rho(Q)Q_s$$

$$- \frac{u\cos k}{2\pi}[f_2(\lambda, B) + f_2(\lambda, -B)]\sigma(B)B_s.$$

Two new functions f_1 and f_2 are defiend as

$$f_1(x, y) = \frac{1}{(u/4)^2 + (y - \sin x)^2}, \tag{B.5}$$

$$f_2(x, y) = \frac{1}{(u/2)^2 + (x - y)^2}, \tag{B.6}$$

and $Q_s \equiv \partial Q / \partial s$, $B_s \equiv \partial B / \partial s$ are determined from the relations

$$\int_{-Q}^{Q} dk \rho_s(k) dk + 2\rho(Q) Q_s = -2\Theta(-u) , \qquad (\text{B.7})$$

$$\int_{-B}^{B} dk \sigma_s(\lambda) d\lambda + 2\sigma(B) B_s = -1 . \qquad (\text{B.8})$$

Appendix C
The chemical potential in the 1D homogeneous Hubbard model from the Bethe-Ansatz solution

The chemical potential for the homogeneous Hubbard model is defiend as

$$\mu(n, s; u) = \frac{\partial}{\partial n} \varepsilon_{\text{GS}}(n, s; u), \tag{C.1}$$

and by equation (3.72) for the ground state energy we have

$$\mu(n, s; u) = \frac{u}{2} \Theta(-u) - 2t \int_{-Q}^{Q} dk \rho_n(k) \cos(k) \tag{C.2}$$
$$- 4t \, Q_n \, \rho(Q) \cos Q \,,$$

where $\rho_n(k) \equiv \partial \rho(k)/\partial n$ and $\sigma_n(\lambda) \equiv \partial \sigma(\lambda)/\partial n$ satisfy the following integral equations [188, 189]:

$$\rho_n(k) = \frac{u \cos k}{4\pi} \int_{-B}^{B} d\lambda \sigma_n(\lambda) f_1(k, \lambda) \tag{C.3}$$
$$+ \frac{u \cos k}{4\pi} [f_1(k, B) + f_1(k, -B)] \sigma(B) B_n \,,$$

$$\sigma_n(\lambda) = \frac{u \cos k}{4\pi} \int_{-Q}^{Q} dk \rho_n(k) f_1(k, \lambda) \tag{C.4}$$
$$- \frac{u \cos k}{2\pi} \int_{-B}^{B} d\lambda' \sigma_n(\lambda') f_2(\lambda, \lambda')$$
$$+ \frac{u \cos k}{4\pi} [f_1(Q, \lambda) + f_1(-Q, \lambda)] \rho(Q) Q_n$$
$$- \frac{u \cos k}{2\pi} [f_2(\lambda, B) + f_2(\lambda, -B)] \sigma(B) B_n \,.$$

f_1 and f_2 are defined in Appendix B by Eqns. (B.5)-(B.6), and $Q_n \equiv \partial Q / \partial n$, $B_n \equiv \partial B / \partial n$ are determined from the relations

$$\int_{-Q}^{Q} dk \rho_n(k) dk + 2\rho(Q) Q_n = \Theta(u),\qquad(\text{C.5})$$

$$\int_{-B}^{B} dk \sigma_n(\lambda) d\lambda + 2\sigma(B) B_n = \frac{1}{2}.\qquad(\text{C.6})$$

References

[1] A. MINGUZZI, S. SUCCI, F. TOSCHI, M. P. TOSI, and P. VI-GNOLO, Phys. Rep. **395**, 223 (2004).

[2] D. S. PETROV, D. M. GANGARDT, and G. V. SHLYAPNIKOV, J. Phys. IV France **116**, 5 (2004).

[3] D. S. JIN, Physics World (April 2002).

[4] M. H. ANDERSON, J. R. ENSHER, M. R. MATTHEWS, C. E. WIEMAN, and E. A. CORNELL, Science **269**, 198 (1995).

[5] K. B. DAVIS, M. O. MEWES, M. R. ANDREWS, N. J. VAN DRUTEN, D. S. DURFEE, D. M. KURN, and W. KETTERLE, Phys. Rev. Lett. **75**, 3969 (1995).

[6] J. T. M. WALRAVEN, E. R. ELIEL, and I. F. SILVERA, Phys. Lett. A **66**, 247 (1978).

[7] T. HÄNSCH and A. SCHAWLOW, Opt. Commun. **13**, 68 (1975).

[8] J. DALIBARD and C. COHEN-TANNOUDJI, J. Opt. Soc. Am. B **6**, 2023 (1989).

[9] A. ASPECT, E. ARIMONDO, R. KAISER, N. VANSTEENKISTE, and C. COHEN-TANNOUDJI, Phys. Rev. Lett **61**, 826 (1988).

[10] M. A. OLSHANII and V. G. MINOGIN, Quantum Opt. **3**, 317 (1991).

[11] J. LAWALL, S. KULIN, B. SAUBAMZEA, N. BIGELOW, M. LEDUC, and C. COHEN-TANNOUDJI, Phys. Rev. Lett. **75**, 4194 (1995).

[12] W. PETRICH, M. H. ANDERSON, J. R. ENSHER, and E. A. CORNELL, Phys. Rev. Lett. **74**, 3352 (1995).

[13] E. A. CORNELL and C. E. WIEMAN, Rev. Mod. Phys. **74**, 875 (2002).

[14] W. KETTERLE, Rev. Mod. Phys. **74**, 1131 (2002).

[15] V. BAGNATO and D. KLEPPNER, Phys. Rev. A **44**, 7439 (1991).

[16] M. ABRAMOWITZ and I. A. STEGAN, *Handbook of Mathematical Function*, Dover Publications, 1970.

[17] W. KETTERLE and N. J. VAN DRUTEN, Phys. Rev. A **54**, 656 (1996).

[18] E. H. LIEB and W. LINIGER, Phys. Rev. **130**, 1605 (1963).

[19] C. N. YANG and C. P. YANG, J. Math. Phys **10**, 1115 (1969).

[20] T. KINOSHITA, T. WENGER, and D. S. WEISS, Science **305**, 1125 (2004).

[21] M. GIRARDEAU, J. Math. Phys **1**, 516 (1960).

[22] A. LENARD, J. Math. phys **5**, 930 (1964).

[23] B. PAREDES, A. WIDERA, V. MURG, O. MANDEL, S. FÖLLING, J. I. CIRAC, G. V. SHLYAPNIKOV, T. W. HÄNSCH, and I. BLOCH, Nature **429**, 277 (2004).

[24] H. MORITZ, T. STÖFERLE, M. KÖHL, and T. ESSLINGER, Phys. Rev. Lett **91**, 250402 (2003a).

[25] D. A. BUTTS and D. S. ROKHSAR, Phys. Rev. A **55**, 4346 (1997).

[26] J. E. THOMAS and M. E. GEHM, American Scientist **92**, 238 (2004).

[27] B. DEMARCO and D. S. JIN, Science **285**, 1703 (1999).

[28] A. G. TRUSCOTT, K. E. STRECKER, W. I. MCALEXANDER, G. B. PARTRIDGE, and R. G. HULET, Science **291**, 2570 (2001a).

[29] F. SCHRECK, L. KHAYKOVICH, K. L. CORWIN, G. FERRARI, T. BOURDEL, J. CUBIZOLLES, and C. SALOMON, Phys. Rev. Lett. **87**, 080403 (2001a).

[30] F. SCHRECK, L. KHAYKOVICH, K. L. CORWIN, G. FERRARI, T. BOURDEL, J. CUBIZOLLES, and C. SALOMON, Phys. Rev. Lett. **87**, 080403 (2001b).

[31] A. G. TRUSCOTT, K. E. STRECKER, W. I. MCALEXANDER, G. B. PARTRIDGE, and R. G. HULET, Science **291**, 2570 (2001b).

[32] M. AMORUSO, I. MECCOLI, A. MINGUZZI, and M. P. TOSI, Eur. Phys. J. D **8**, 361 (2000).

[33] S. D. GENSEMER and D. S. JIN, Phys. Rev. Lett. **87**, 173201 (2001).

[34] B. DEMARCO and D. S. JIN, Phys. Rev. Lett. **88**, 040405 (2002).

[35] B. DEMARCO, S. B. PAPP, and D. S. JIN, Phys. Rev. Lett. **86**, 5409 (2001).

[36] M. OLSHANII, Phys. Rev. Lett. **81**, 938 (1998).

[37] K. HUANG, *Statistical Mechanics*, Wiley, 1987.

[38] M. G. MOORE, T. BERGEMAN, and M. OLSHANII, J. Phys. IV France **116**, 69 (2004).

[39] T. BERGEMAN, M. G. MOORE, and M. OLSHANII, Phys. Rev. Lett **91**, 163201 (2003).

[40] H. MORITZ, T. STÖFERLE, K. GÜNTER, M. KÖHL, and T. ESSLINGER, Phys. Rev. Lett **94**, 210401 (2005).

[41] G. F. GIULIANI and G. VIGNALE, *Quantum Theory of the Electron Liquid*, Cambridge University Press, Cambridge, 2005.

[42] J. VOIT, Rep. Progr. Phys. **58**, 977 (1995).

[43] H. J. SCHULZ, G. CUNIBERTI, and P. PIERI, IN *Field Theories for Low-Dimensional Condensed Matter systems*, EDITED BY G. MORANDI, P. SODANO, A. TAGLIACOZZO, and V. TOGNETTI, Springer, Berlin, 2000.

[44] T. GIAMARCHI, *Quantum Physics in One Dimension*, Clarendon Press, Oxford, 2004.

[45] D. W. WANG, A. J. MILLIS, and S. DAS SARMA, Phys. Rev. Lett. **85**, 4570 (2000).

[46] S. TOMONAGA, Progr. Theor. Phys. **5**, 544 (1950).

[47] J. M. LUTTINGER, J. Math. Phys. **4**, 1154 (1963).

[48] D. C. MATTIS and E. H. LIEB, J. Math. Phys. **6**, 304 (1965).

[49] A. LUTHER and I. PESCHEL, Phys. Rev. B **9**, 2911 (1974).

[50] F. D. M. HALDANE, J. Phys. C **14**, 2585 (1981).

[51] A. LUTHER and V. J. EMERY, Phys. Rev. Lett **33**, 589 (1974).

[52] A. SEIDEL and D.-H. LEE, Phys. Rev. B **71**, 045113 (2005).

[53] A. SEIDEL and D.-H. LEE, Phys. Rev. Lett **93**, 046401 (2004).

[54] J. W. TURKSTRA, PhD Thesis, Rijksuniversiteit Groningen, Netherland (2001).

[55] S. N. ATUNOV, Eur. Phys. J. D **13**, 71 (2001).

[56] M. GREINER, PhD Thesis, Ludwig-Maximilians-Universität München, Germany (2003).

[57] E. L. RAAB, M. PRENTISS, A. CABLE, S. CHU, and D. E. PRITCHARD, Phys. Rev. Lett. **59**, 2631 (1987).

[58] H. MORITZ, T. STÖFERLE, M. KÖHL, and T. ESSLINGER, Phys. Rev. Lett. **91**, 250402 (2003b).

[59] D. JAKSCH, PhD Thesis, Insbruck, Austria (1999).

[60] M. GREINER, O. MANDEL, T. ESSLINGER, T. HÄNSCH, and I. BLOCH, Nature **415**, 39 (2002).

[61] M. P. A. FISHER, P. B. WEICHMAN, G. GRINSTEIN, and D. S. FISHER, Phys. Rev. B. **40**, 546 (1989).

[62] D. JAKSCH, C. BRUDER, J. I. CIRAC, C. W. GARDINER, and P. ZOLLER, Phys. Rev. Lett. **81**, 3108 (1998).

[63] J. I. CIRAC and P. ZOLLER, Science **301**, 176 (2003).

[64] I. BLOCH, J. Phys. B: At. Mol. Opt. Phys. **38**, S629 (2005).

[65] N. ELSTNER and H. MONIEN, Phys. Rev. B. **59**, 12184 (1999).

[66] C. A. REGAL, C. TICKNOR, J. L. BOHN, and D. S. JIN, Nature **424**, 47 (2003).

[67] J. CUBIZOLLES, T. BOURDEL, S. J. J. M. F. KOKKELMANS, G. V. SHLYAPNIKOV, and C. SALOMON, Phys. Rev. Lett. **91**, 240401 (2003).

[68] S. JOCHIM, M. BARTENSTEIN, A. ALTMEYER, G. HENDL, C. CHIN, J. H. DENSCHLAG, and R. GRIMM, Phys. Rev. Lett. **91**, 240402 (2003a).

[69] M. W. ZWIERLEIN, C. A. STAN, C. H. SCHUNCK, S. M. F. RAUPACH, S. GUPTA, Z. HADZIBABIC, and W. KETTERLE, Phys. Rev. Lett. **91**, 250401 (2003).

[70] M. GREINER, C. A. REGAL, and D. S. JIN, Nature **426**, 537 (2003).

[71] S. JOCHIM, M. BARTENSTEIN, A. ALTMEYER, G. HENDL, S. RIEDL, C. CHIN, J. H. DENSCHLAG, and R. GRIMM, Science **302**, 2101 (2003b).

[72] M. GREINER AND C.A. REGAL and D.S. JIN, cond-mat/ 0502539.

[73] M. TINKHAM, *Introduction to Superconductivity*, Dover Publication, INC, 1996.

[74] C. A. REGAL and D. S. JIN, Phys. Rev. Lett. **90**, 230404 (2003).

[75] J. YIN, Phys. Rep. **430**, 1 (2006).

[76] C. A. REGAL, M. GREINER, and D. S. JIN, Phys. Rev. Lett. **92**, 040403 (2004).

[77] M. HOLLAND, S. J. J. M. F. KOKKELMANS, M. L. CHIOFALO, and R. WALSER, Phys. Rev. Lett. **87**, 120406 (2001).

[78] M. W. ZWIERLEIN, C. A. STAN, C. H. SCHUNCK, S. M. F. RAUPACH, A. J. KERMAN, and W. KETTERLE, Phys. Rev. Lett. **92**, 120403 (2004).

[79] M. R. BAKHTIARI, P. VIGNOLO, and M. P. TOSI, Physica E **28**, 385 (2005).

[80] M. R. BAKHTIARI, P. VIGNOLO, and M. P. TOSI, Physica E **33**, 222 (2006).

[81] P. VIGNOLO, M. R. BAKHTIARI, and M. P. TOSI, IN *New Developments in Condensed Matter Physics*, Nova Publisher, 2006.

[82] R. LANDAUER, Phil. Mag. **21**, 863 (1970).

[83] E. N. ECONOMOU, *Green's Functions in Quantum Physics*, Springer: Berlin, 1983.

[84] L. JACAK, P. HAWRYLAK, and A. WÒJS, *Quantum Dots*, Springer: Berlin, 1998.

[85] T. CHAKRABORTY, *Quantum Dots: A Survey of the Properties of Artificial Atoms*, Elsevier: Amsterdam, 1999.

[86] M. TEWS, Ann. Phys. (Leipzig) **13**, 249 (2004).

[87] V. MOLDOVEANU, A. ALDEA, A. MANOLESCU, and M. NITA, Phys. Rev. B **63**, 045301 (2001).
[88] E. LOUIS and J. A. VERGES, Phys. Rev. B **63**, 115310 (2001).
[89] M. E. TORIO, K. HALLBERG, A. H. CECCATTO, and C. R. PROETTO, Phys. Rev. B **65**, 085302 (2002).
[90] G. KIRCZENOW, Phys. Rev. B **46**, 1439 (1992).
[91] V. MOLDOVEANU, A. ALDEA, and B. TANATAR, Phys. Rev. B **70**, 085303 (2004).
[92] A. DORN, T. IHN, K. ENSSLIN, W. WEGSCHEIDER, and M. BICHLER, Phys. Rev. B **70**, 205306 (2004).
[93] M. MARDANI and K. ESFARJANI, Physica E **25**, 119 (2004).
[94] M. MARDANI and K. ESFARJANI, Physica E **27**, 227 (2005).
[95] M. B. DAHAN, E. PEIK, J. REICHEL, Y. CASTIN, and C. SALOMON, Phys. Rev. Lett **76**, 4508 (1996).
[96] B. P. ANDERSON and M. A. KASEVICH, Science **282**, 1686 (1998).
[97] C. F. BHARUCHA, K. W. MADISON, P. R. MORROW, S. R. WILKINSON, B. SUNDARAM, and M. G. RAIZEN, Phys. Rev. A **55**, 857 (1997).
[98] S. BURGER, F. S. CATALIOTTI, C. FORT, F. MINARDI, M. INGUSCIO, M. L. CHIOFALO, and M. P. TOSI, Phys. Rev. Lett **86**, 4447 (2001).
[99] F. S. CATALIOTTI, S. BURGER, C. FORT, P. MADDALONI, F. MINARDI, A. TROMBETTONI, A. SMERZI, and M. INGUSCIO, Science **293**, 843 (2001).
[100] J. STEINHAUER, N. KATZ, R. OZERI, N. DAVIDSON, C. TOZZO, and F. DALFOVO, Phys. Rev. Lett **90**, 060404 (2003).
[101] G. ROATI, E. DE MIRANDES, F. FERLAINO, H. OTT, G. MODUGNO, and M. INGUSCIO, Phys. Rev. Lett **92**, 230402 (2004).
[102] P. GIANNOZZI, G. GROSSO, S. MORONI, and G. P. PARRAVICINI, Appl. Numer. Math **4**, 273 (1988).
[103] G. GROSSO and G. P. PARRAVICINI, Adv. Chem. Phys. **62**, 81 (1986).
[104] K. BERG-SÖRENSEN and K. MÖLMER, Phys. Rev. A **58**, 1480 (1998).
[105] M. L. CHIOFALO, M. POLINI, and M. P. TOSI, Eur. Phys. J. D **11**, 371 (2000).
[106] M. L. CHIOFALO, S. SUCCI, and M. P. TOSI, Phys. Lett. A **260**, 86 (1999).
[107] M. L. CHIOFALO and M. P. TOSI, EuroPhys. Lett **56**, 326 (2001).
[108] P. VIGNOLO, Z. AKDENIZ, and M. P. TOSI, J. Phys. B **36**, 4535 (2003).

[109] Y. EKSIOGLU, P. VIGNOLO, and M. P. TOSI, Laser Phys. **15**, 356 (2005).

[110] Y. EKSIOGLU, P. VIGNOLO, and M. P. TOSI, Optics Commun. **243**, 175 (2004).

[111] P. D. KIRKMAN and J. B. PENDRY, J. Phys. C **17**, 4327 (1984).

[112] R. FARCHIONI, G. GROSSO, and G. P. PARRAVICINI, Phys. Rev. B **45**, 6383 (1992).

[113] R. FARCHIONI, P. VIGNOLO, and G. GROSSO, Phys. Rev. B **60**, 15705 (1999).

[114] R. FARCHIONI, G. GROSSO, and P. VIGNOLO, Phys. Rev. B **62**, 12565 (2000).

[115] X. XU, K. KIM, W. JHE, and N. KWON, Phys. Rev. A **63**, 063401 (2001).

[116] C. ZENER, Proc. R. Soc. London A **145**, 523 (1934).

[117] M. Paulsson, cond-mat/0210519.

[118] J. C. SLATER, Phys. Rev. **87**, 807 (1952).

[119] L. SALASNICH, A. PAROLA, and L. REATTO, Phys. Rev. A **65**, 043614 (2002).

[120] G. ROATI, F. RIBOLI, G. MODUGNO, and M. INGUSCIO, Phys. Rev. Lett **89**, 150403 (2002).

[121] J. J. SAKURAI, *Modern Quantum Mechanics*, Addison Wesley: Reading, 1994.

[122] E. H. LIEB and F. Y. WU, Phys. Rev. Lett. **20**, 1445 (1968).

[123] P. HOHENBERG and W. KOHN, Phys. Rev. **136**, B864 (1964).

[124] W. KOHN and L. J. SHAM, Phys. Rev. **140**, A1133 (1965).

[125] W. KOHN, Rev. Mod. Phys. **71**, 1253 (1999).

[126] R. M. DREIZLER and E. K. U. GROSS, *Density Functional Theory*, Springer-Verlag, 1990.

[127] K. CAPELLE, *A bird's-eye view of density-functional theory* (COND-MAT/0211443).

[128] W. KOHN, A. D. BECKE, and R. G. PARR, J. Phys. Chem. **100**, 12974 (1996).

[129] O. GUNNARSSON and B. I. LUNDQVIST, Phys. Rev. B **13**, 4274 (1976).

[130] K. CAPELLE and G. VIGNALE, Phys. Rev. Lett. **86**, 5546 (2001).

[131] L. J. SHAM and M. SCHLÜTER, Phys. Rev. Lett. **51**, 1888 (1983).

[132] J. P. PERDEW and M. LEVY, Phys. Rev. Lett. **51**, 1884 (1983).

[133] R. W. GODBY, M. SCHLÜTER, and L. J. SHAM, Phys. Rev. Lett. **56**, 2415 (1986).

[134] O. GUNNARSSON and K. SCHÖNHAMMER, Phys. Rev. Lett. **56**, 1968 (1986).

[135] K. SCHÖNHAMMER and O. GUNNARSSON, J. Phys. C **20**, 3675 (1987).

[136] J. HUBBARD, Proc. Roy. Soc. (London) A **276**, 238 (1963).

[137] J. HUBBARD, Proc. Roy. Soc. (London) A **277**, 237 (1964).

[138] Y. REN and P. W. ANDERSON, Phys. Rev. B **48**, 16662 (1993).

[139] E. H. Lieb and F. Y. Wu, cond-mat/0207529.

[140] K. SCHÖNHAMMER and O. GUNNARSSON, Phys. Rev. B **37**, 3128 (1988).

[141] K. SCHÖNHAMMER, O. GUNNARSSON, and R. M. NOACK, Phys. Rev. B **52**, 2504 (1995).

[142] H. J. SCHULZ, Phys. Rev. Lett. **64**, 2831 (1990).

[143] T. GIAMARCHI, Physica B **230-232**, 975 (1997).

[144] F. H. L. ESSLER, H. FRAHM, F. GÖHMANN, A. KLÜMPER, and V. E. KOREPIN, *The One-Dimensional Hubbard Model*, Cambridge University Press, 2005.

[145] H. J. SCHULZ, IN *Proceedings of Les Houches Summer School LXI*, EDITED BY E. AKKERMANS, G. MONTAMBAUX, J. L. PICHARD, and J. ZINN-JUSTIN, Elsevier, 1995, p. 533, (cond-mat/9503150).

[146] F. D. M. HALDANE, Phys. Rev. Lett. **45**, 1358 (1980).

[147] G. XIANLONG, M. POLINI, M. P. TOSI, V. L. CAMPO, K. CAPELLE, and M. RIGOL, Phys. Rev. B **73**, 165120 (2006a).

[148] G. XIANLONG, M. POLINI, B. TANATAR, and M. P. TOSI, Phys. Rev. B **73**, 161103 (2006b).

[149] N. A. LIMA, M. F. SILVA, L. N. OLIVEIRA, and K. CAPELLE, Phys. Rev. Lett. **90**, 146402 (2003).

[150] GAO XIANLONG, M. RIZZI, M. POLINI, R. FAZIO, M. P. TOSI, JR. V. L. CAMPO, and K. CAPELLE, cond-mat/0609346.

[151] A. I. LARKIN and Y. N. OVCHINNIKOV, Zh. Eksp. Teor. Fiz **47**, 1136 (1964), [Sov. Phys. JETP 20,762, (1965).

[152] P. FULDE and R. A. FERRELL, Phys. Rev. **135**, A550 (1964).

[153] R. CASALBUONI and G. NARDULLI, Rev. Mod. Phys. **76**, 263 (2004).

[154] M. ALFORD, J. A. BOWERS, and K. RAJAGOPAL, Phys. Rev. D **63**, 074016 (2001).

[155] K. YANG, In: *Pairing beyond BCS Theory in Fermionic Systems*, M. Alford, J. Clark, and A. Sedrakian (eds.), World Scientific, 2006.

[156] A. M. CLOGSTON, Phys. Rev. Lett. **9**, 266 (1962).

[157] R. COMBESCOT, Europhys. Lett. **55**, 150 (2001).

[158] W. V. LIU and F. WILCZEK, Phys. Rev. Lett. **90**, 047002 (2003).

[159] P. F. BEDAQUE, H. CALDAS, and G. RUPAK, Phys. Rev. Lett. **91**, 247002 (2003).

[160] S.-T. WU and S. YIP, Phys. Rev. A **67**, 053603 (2003).

[161] M. M. FORBES, E. GUBANKOVA, W. V. LIU, and F. WILCZEK, Phys. Rev. Letters **94**, 017001 (2005).

[162] T. MIZUSHIMA, K. MACHIDA, and M. ICHIOKA, Phys. Rev. Lett. **94**, 060404 (2005).

[163] K. MACHIDA and H. NAKANISHI, Phys. Rev. B **30**, 122 (1984).

[164] K. YANG, Phys. Rev. B **63**, 140511 (2001).

[165] K. YANG, Phys. Rev. Lett. **95**, 218903 (2005).

[166] E. ALTMAN, E. DEMLER, and M. D. LUKIN, Phys. Rev. A **70**, 013603 (2004).

[167] M. GREINER, C. A. REGAL, J. T. STEWART, and D. S. JIN, Phys. Rev. Lett. **94**, 110401 (2005).

[168] K. YANG and D. F. AGTERBERG, Phys. Rev. Lett. **84**, 4970 (2000).

[169] M. W. ZWIERLEIN, A. SCHIROTZEK, C. H. SCHUNCK, and W. KETTERLE, Science **311**, 492 (2006a).

[170] G. B. PARTRIDGE, W. LI, R. I. KAMAR, Y. LIAO, and R. G. HULET, Science **311**, 503 (2006).

[171] M. W. ZWIERLEIN, C. H. SCHUNCK, A. SCHIROTZEK, and W. KETTERLE, Nature **442**, 54 (2006b).

[172] Y. SHIN, M. W. ZWIERLEIN, C. H. SCHUNCK, A. SCHIROTZEK, and W. KETTERLE, Phys. Rev. Lett. **97**, 030401 (2006).

[173] Y.-P. SHIM, R.A. DUINE, and A. H. MACDONALD, cond-mat/0608255.

[174] M. W. ZWIERLEIN, J. R. ABO-SHAEER, A. SCHIROTZEK, C. H. SCHUNCK, and W. KETTERLE, Nature **435**, 1047 (2005).

[175] D. E. SHEEHY and L. RADZIHOVSKY, Phys. Rev. Lett. **96**, 060401 (2006).

[176] P. CASTORINA, M. GRASSO, M. OERTEL, M. URBAN, and D. ZAPPALA, Phys. Rev. A **72**, 025601 (2005).

[177] J. E. THOMAS, Nature **442**, 32 (2006).

[178] L. N. OLIVEIRA, E. K. U. GROSS, and W. KOHN, Phys. Rev. Lett. **60**, 2430 (1988).

[179] P.-G. DE GENNES, *Superconductivity of Metals and Alloys*, Benjamin, New York, 1966.

[180] I. AFFLECK, J.-S. CAUX, and A. M. ZAGOSKIN, Phys. Rev. B **62**, 1433 (2000).

[181] Q. WANG, H.-Y. CHEN, C.-R. HU, and C. S. TING, Phys. Rev. Lett. **96**, 117006 (2006).

[182] H.-Y. CHEN and C. S. TING, Phys. Rev. B **71**, 220510 (2005).

[183] J. KINNUNEN, M. RODRIGUEZ, and P. TÖRMÄ, Science **305**, 1131 (2004).

[184] K. MACHIDA, T. MIZUSHIMA, and M. ICHIOKA, Phys. Rev. Lett **97**, 120407 (2006).

[185] J. KINNUNEN, L. M. JENSEN, and P. TÖRMÄ, Phys. Rev. Lett. **96**, 110403 (2006).

[186] R. A. PINTO, M. RODRIGUEZ, J. A. GONZÁLES, and E. MEDINA, Phys. Lett. A **341**, 101 (2005).

[187] M. BORN and E. WOLF, *Principles of Optics*, Pergamon: London, 1959.

[188] C. YANG, A. N. KOCHARIAN, and Y. L. CHIANG, J. Phys.: Condens. matter **12**, 7433 (2000).

[189] A. N. KOCHARIAN, C. YANG, and Y. L. CHIANG, Phys. Rev. B **59**, 7458 (1999).

THESES

This series gathers a selection of outstanding Ph.D. theses defended at the Scuola Normale Superiore since 1992.

Published volumes

1. F. COSTANTINO, *Shadows and Branched Shadows of 3 and 4-Manifolds*, 2005. ISBN 88-7642-154-8

2. S. FRANCAVIGLIA, *Hyperbolicity Equations for Cusped 3-Manifolds and Volume-Rigidity of Representations*, 2005. ISBN 88-7642-167-x

3. E. SINIBALDI, *Implicit Preconditioned Numerical Schemes for the Simulation of Three-Dimensional Barotropic Flows*, 2007. ISBN 978-88-7642-310-9

4. F. SANTAMBROGIO, *Variational Problems in Transport Theory with Mass Concentration*, 2007. ISBN 978-88-7642-312-3

5. M. R. BAKHTIARI, *Quantum Gases in Quasi-One-Dimensional Arrays*, 2007. ISBN 978-88-7642-319-2

Volumes published earlier

H.Y. FUJITA, *Equations de Navier-Stokes stochastiques non homogènes et applications*, 1992.

G. GAMBERINI, *The minimal supersymmetric standard model and its phenomenological implications*, 1993. ISBN 978-88-7642-274-4

C. DE FABRITIIS, *Actions of Holomorphic Maps on Spaces of Holomorphic Functions*, 1994. ISBN 978-88-7642-275-1

C. PETRONIO, *Standard Spines and 3-Manifolds*, 1995. ISBN 978-88-7642-256-0

I. DAMIANI, *Untwisted Affine Quantum Algebras: the Highest Coefficient of* det H_η *and the Center at Odd Roots of 1*, 1996. ISBN 978-88-7642-285-0

M. MANETTI, *Degenerations of Algebraic Surfaces and Applications to Moduli Problems*, 1996. ISBN 978-88-7642-277-5

F. CEI, *Search for Neutrinos from Stellar Gravitational Collapse with the MACRO Experiment at Gran Sasso*, 1996. ISBN 978-88-7642-284-3

A. SHLAPUNOV, *Green's Integrals and Their Applications to Elliptic Systems*, 1996. ISBN 978-88-7642-270-6

R. TAURASO, *Periodic Points for Expanding Maps and for Their Extensions*, 1996. ISBN 978-88-7642-271-3

Y. BOZZI, *A study on the activity-dependent expression of neurotrophic factors in the rat visual system*, 1997. ISBN 978-88-7642-272-0

M.L. CHIOFALO, *Screening effects in bipolaron theory and high-temperature superconductivity*, 1997. ISBN 978-88-7642-279-9

D.M. CARLUCCI, *On Spin Glass Theory Beyond Mean Field*, 1998. ISBN 978-88-7642-276-4

G. LENZI, *The MU-calculus and the Hierarchy Problem*, 1998. ISBN 978-88-7642-283-6

R. SCOGNAMILLO, *Principal G-bundles and abelian varieties: the Hitchin system*, 1998. ISBN 978-88-7642-281-2

G. ASCOLI, *Biochemical and spectroscopic characterization of CP20, a protein involved in synaptic plasticity mechanism*, 1998. ISBN 978-88-7642-273-7

F. PISTOLESI, *Evolution from BCS Superconductivity to Bose-Einstein Condensation and Infrared Behavior of the Bosonic Limit*, 1998. ISBN 978-88-7642-282-9

L. PILO, *Chern-Simons Field Theory and Invariants of 3-Manifolds*, 1999. ISBN 978-88-7642-278-2

P. ASCHIERI, *On the Geometry of Inhomogeneous Quantum Groups*, 1999. ISBN 978-88-7642-261-4

S. CONTI, *Ground state properties and excitation spectrum of correlated electron systems*, 1999. ISBN 978-88-7642-269-0

G. GAIFFI, *De Concini-Procesi models of arrangements and symmetric group actions*, 1999. ISBN 978-88-7642-289-8

N. DONATO, *Search for neutrino oscillations in a long baseline experiment at the Chooz nuclear reactors*, 1999. ISBN 978-88-7642-288-1

R. CHIRIVÌ, *LS algebras and Schubert varieties*, 2003. ISBN 978-88-7642-287-4

V. MAGNANI, *Elements of Geometric Measure Theory on Sub-Riemannian Groups*, 2003. ISBN 88-7642-152-1

F.M. ROSSI, *A Study on Nerve Growth Factor (NGF) Receptor Expression in the Rat Visual Cortex: Possible Sites and Mechanisms of NGF Action in Cortical Plasticity*, 2004. ISBN 978-88-7642-280-5

G. PINTACUDA, *NMR and NIR-CD of Lanthanide Complexes*, 2004. ISBN 88-7642-143-2

Fotocomposizione "CompoMat" Loc. Braccone, 02040 Configni (RI) Italy
Finito di stampare nel mese di dicembre 2007 presso
BRAILLE-GAMMA s.r.l. Cittaducale (RI)